PROBLEM SOLVING IN ANALYTICAL BIOCHEMISTRY

PROBLEM SOLVING IN ANALYTICAL BIOCHEMISTRY

David J. Holme and Hazel Peck

Division of Biomedical Sciences, Sheffield Hallam University

Longman Scientific & Technical
Longman Group UK Limited
Longman House, Burnt Mill, Harlow
Essex CM20 2JE England
and Associated Companies throughout the world.

Copublished in the United States with
John Wiley & Sons Inc., 605 Third Avenue, New York, NY 10158

First published 1994

British Library Cataloguing in Publication Data
A catalogue record for this title is available from the British Library

ISBN: 0582 - 22710-0

Library of Congress Cataloging-in-Publication Data

Holme, David J. (David James), 1934-
Problem Solving in analytical biochemistry / David J. Holme and Hazel Peck
p. cm.
ISBN 0-470-23378-8
1. Analytical biochemistry -- Problems, exercises, etc. I. Peck, Hazel, 1946- II. Title.
QP519.7.H64 1993 Suppl.
574.19'285'076--dc20 93-49460
CIP

Printed in Malaysia

CONTENTS

Problems marked with an asterisk are unworked test questions.

PREFACE

Analytical Biochemistry is essentially a practical subject, but it has often been taught with an emphasis on the theoretical aspects, with less attention being paid to the application of this knowledge in solving specific analytical problems. This deficiency is now being recognised and the need for students to be able to apply their knowledge in method design and assessment is being addressed; a fact reflected in the increasing availability of computer-based simulations of analytical situations.

We have increasingly introduced problem solving into our own teaching and appreciated the benefits in terms of student motivation and the significant increase in their understanding of the principles involved. It was with this in mind that we planned to include a series of data-based problems in the second edition of our textbook *Analytical Biochemistry* (Longman, 1993). However, it soon became apparent that these problems justified a separate volume; hence this publication was born.

During the preparation of this book we discovered that some problems, although excellent in tutorial situations, did not lend themselves to a written question with a short, unambiguous answer. We have tried, therefore, to restrict ourselves to problems for which the outcome is reasonably straightforward. Some answers, however, do require an explanation and in these situations we have included a brief discussion.

For convenience all the problems are presented in the same format. They are grouped into subject areas, and a brief synopsis of the basic scientific information required to attempt to answer the questions precedes each section. As the book is primarily designed for students working on their own, answers are supplied in the majority of cases to enable the reader to check their conclusions. However to permit the use of the book in teaching situations a number of the problems are left unworked.

David J. Holme
Hazel Peck
Sheffield Hallam University

February 1993

ABOUT THIS BOOK

The book consists of forty problems which are grouped into nine subject areas, each section being introduced by a brief review of the essential background scientific information relevant to the problems that follow. If further information is required, it should be sought in more comprehensive text books on analytical methods and, to assist in this way, all the problems carry references to specific sections in *Analytical Biochemistry* (Holme & Peck, Longman, 1993, 2nd edition).

Each problem is posed in the same manner, and guidance on solving the problem and the worked answers are included for the majority of the problems. These sections should only be consulted when necessary. The remainder of the problems, headed 'Test Questions', are unworked and do not include these sections. They may be useful for class or course work applications.

QUESTIONS DESIGN

The problems are all presented in a similar format under the headed sections which are explained below. The section entitled 'Solving the problem' should not be consulted until an attempt has been made to answer the question unaided. The unworked questions include neither this nor the 'Answer' section.

INTRODUCTION

In this section the background knowledge and scientific concepts required to be able to solve the problem are stated. Reference is made to other sections in the book where useful basic information may be found. Specific references to the relevant sections in the text book *Analytical Biochemistry* are also given where more detailed information is available.

ANALYTICAL PROBLEM

A specific analytical problem is outlined that is designed to be representative of many similar situations which occur in analytical biochemistry laboratories. Specific questions are posed which need to be answered in order to solve the problem, using the subsequent data and information provided. These questions indicate what would need to be considered when resolving similar problems.

INVESTIGATIONS

The investigations which were undertaken to solve the problem are given. They also indicate what general experimental approach may be appropriate to resolve comparable analytical problems.

DATA

The results of the investigations are presented as numerical data, spectra, chromatograms, etc. as appropriate. It is intended that measurements and readings should be taken from graphs, recorder traces, etc. provided in the figures in order to answer the questions posed.

SOLVING THE PROBLEM

This section should only be consulted *AFTER* an attempt had been made at analysing the data and answering the questions. It is presented as the sequence of steps which need to be taken to resolve the problem.

ANSWERS

The answers for each of the worked questions and, where relevant, a discussion of the significance of the results together with comments on other aspects raised by the questions or the data, are presented together at the end of the book. Problems are based on authentic data but since diagrams have been simplified for clarity, numerical answers derived from them may differ slightly from published data.

1

ASSESSMENT OF QUANTITATIVE METHODS

PRECISION

The precision, or reproducibility, of a method is defined as the closeness of a number of replicate measurements (n). It is most conveniently expressed in terms of the standard deviation (s) of a large number (n) of replicate determinations (i.e. greater than thirty).

$$s = \sqrt{\frac{\sum x^2 - \frac{(\sum x)^2}{n}}{n}}$$ where x is an individual measurement.

The use of coefficient of variation (V) is a convenient way of expressing the variation which occurs as a percentage of the mean value ($\overline{x}$). The significance of a percentage is easier to appreciate than a standard deviation value, which is purely a mathematical term.

$$V = \frac{s}{\overline{x}} \times 100\%$$

For some analytical methods precision may vary depending upon the concentration of the analyte. Therefore, before any quantitative method can be reliably used, its precision must be assessed over the working range.

A comparison of the precision given by the two methods may assist in the choice of a method for routine use. Statistical comparison of values for the standard deviation using the 'F' test may be used to compare not only different methods but also the results from different analysts or laboratories.

The equation for the 'F' test is given below and if there is no significant difference between the standard deviation of the two sets of data, the value for F will be unity. If there is significant difference between the data, the calculated value for F will be greater than the expected (critical) value. This value can be found in published statistical tables.

$$F_{\phi}^{\phi} = \frac{s_1^2}{s_2^2} = \frac{\text{largest variance estimate}}{\text{smallest variance estimate}}$$

where ϕ is the number of degrees of freedom.

ACCURACY

Accuracy is defined as the closeness of the mean value of replicate analyses to the true value for the sample. It is often only possible to assess the accuracy of a method relative to another which, for one reason or another, is assumed to give a true mean value. This can be done by comparing the means of replicate analyses by the two methods using the '*t*' test.

The equation for the '*t*' test is given below. If the calculated value for '*t*' is greater than the expected or critical value found in published statistical tables, then it can be assumed that there is a significant difference between the two mean values.

$$t_{\phi} = \frac{\overline{x}_1 - \overline{x}_2}{\sqrt{\dfrac{\sum(x_1 - \overline{x}_1)^2 + \sum(x_2 - \overline{x}_2)^2}{n(n-1)}}}$$

where $\overline{x}_1$ and $\overline{x}_2$ are the mean values and x_1 and x_2 are the individual measurements by methods 1 and 2 respectively.

The method under study is usually compared with an accepted reference method or, if one is not available, with a method which relies on an entirely different principle.

If a method gives a mean value which differs from the accepted true value, the method is said to show a bias.

SENSITIVITY

The sensitivity of a method is defined as the ability to detect small amounts of the test substance. It can be expressed, for example, as the smallest reading after zero that can be consistently detected and measured. Alternatively it may be quoted in terms of the slope of the calibration graph.

SPECIFICITY

Specificity is the ability to detect only the test substance. Lack of specificity will result in false positives if the method is qualitative, and in quantitative results which are greater than they should be, i.e. positively biased.

QUALITY ASSURANCE

In order to produce reliable results, all methods should be carefully designed and their precision and accuracy must be determined and monitored. A system in which control samples (i.e. samples for which the value is known) are analysed along with the test samples should be initiated and the results monitored in order to detect deviation from the required levels of accuracy. Control samples, whether they be from a commercial source or prepared in the laboratory, should resemble the composition of the test samples as closely as possible, and their mean and standard deviation values should normally be known.

It is often helpful to record the results of control samples in a visual manner. The most frequently used graphical method is that known as the Levey–Jennings or Shewart plot in which the individual results are plotted on a chart that incorporates the acceptable limits. These are often set at ±2s and ±3s which approximate to 95 per cent and 99 per cent confidence limits, respectively. Failure of a control result to fall within the specified limits would suggest that all the results in the batch are also wrong and that the method should be investigated before the test results can be accepted.

An alternative visual method of quality assurance is the Cusum plot. The cumulative difference between the control sample result actually obtained and its previously determined mean value is plotted against the analysis date, or batch number. A definite change in the slope of the line indicates a change in the accuracy of the method. However, this is only apparent in retrospect and the method is less useful than the Levey–Jennings plot for giving instant information regarding poor performance.

PROBLEM 1 A COMPARISON OF THE ANALYTICAL PERFORMANCE OF TWO METHODS

INTRODUCTION

Before any new method is introduced it is necessary to determine whether it is capable of producing results which are as reliable as those produced by an alternative method. The concepts of accuracy and precision are important, as is the ability to use statistical tests to assess the significance of any differences between numerical values. Refer to the 'Assessment of Quantitative Methods' (p. 1) for background information, and also to *Analytical Biochemistry* (§ 1.2.2, 1.2.3).

ANALYTICAL PROBLEM

In a routine analytical laboratory it is proposed to replace an existing, reliable quantitative method (A) by a newly developed, rapid method (B) and the performance of the proposed method must be assessed.

The following question needs to be answered:

■ How does the precision, sensitivity and accuracy of Method B compare with Method A?

INVESTIGATIONS

The following samples were analysed using both methods:

- a series of standard solutions;

- replicate analyses of a single sample.

DATA

	Absorbance	
	Method A	*Method B*
Standards (mg l^{-1})		
5	0.12	0.08
10	0.24	0.17
15	0.36	0.25
20	0.48	0.34
Replicate analyses	0.25	0.24
	0.28	0.29
	0.31	0.32
	0.34	0.24
	0.38	0.30
	0.33	0.24
	0.37	0.32
	0.32	0.26
	0.33	0.29
	0.32	0.32

The critical value for F (with 9 degrees of freedom) is 3.05 and for t (with 18 degrees of freedom) is 2.10.

For simplicity only ten replicate results are given; in practice many more replicates would need to be analysed for reliable conclusions to be drawn.

SOLVING THE PROBLEM

Only refer to this section when you have made an attempt to answer the question.

1. Draw a calibration graph for each method and use this to determine the concentration of the analyte in all the samples.
2. Using the results from the replicate analyses:
   (a) calculate the mean value, standard deviation and coefficient of variation for each method;
   (b) statistically compare the precision of the new method with that of the original method;
   (c) statistically compare the accuracy of the two methods.
3. Calculate the slope of the calibration line for each method.

PROBLEM 2 THE USE OF QUALITY CONTROL CHARTS

INTRODUCTION

Data from quality control samples is usually presented visually on a chart, the most frequently used being the Shewart (Levey–Jennings) and the Cusum plots. Although both of these methods of visual representation give valuable information regarding the quality of the results produced, they each have particular relevance to specific situations. Refer to 'Assessment of Quantitative Methods' (p. 1) for background information and also to *Analytical Biochemistry* (§ 1.2.3).

ANALYTICAL PROBLEM

An internal quality assurance programme has been introduced into a laboratory to assist the technical manager in making decisions about the accuracy of each batch of analyses. The most acceptable method of plotting the quality control results for this purpose needs to be established.

The following question needs to be answered:

- Does the Shewart or the Cusum plot give the most appropriate information?

INVESTIGATIONS

Over a period of time fifty replicate analyses of the control sample were performed to determine the 'between batch' mean and standard deviation.

Quality control samples were analysed with twenty consecutive batches of test samples.

DATA

Sample batch number	*Quality control sample* (μmol l^{-1})
1	139
2	140
3	139
4	139
5	140
6	142
7	139
8	140
9	140
10	139
11	140

(continued)

12	138
13	137
14	137
15	138
16	137
17	138
18	137
19	139
20	137

The mean and standard deviation of the quality control sample for the 50 replicates was 138 μmol l^{-1} and 1.2 μmol l^{-1}, respectively.

SOLVING THE PROBLEM

Only refer to this section when you have made an attempt to answer the question.

1. From the data, construct the two types of plot and decide upon the acceptable limits for the Shewart plot.
2. Note any quality control results which lie outside the acceptable limits on the Shewart plot. Observe whether this is also apparent from the Cusum plot.

PROBLEM 3 AN INVESTIGATION INTO THE SPECIFICITY OF AN IMMUNOASSAY

INTRODUCTION

Immunoassay provides very novel methods for the assay of many substances. However, similarities between the analyte and related compounds can sometimes result in cross-reactions and a reduction in specificity, the implications of which can be very significant. Refer to the 'Assessment of Quantitative Methods' (p. 1) for background information. The principles of immunoassay are important (refer to *Analytical Biochemistry*, § 7.4) but the quantitative aspects involving the use of calibration graphs which are required for this question are common to many different methods.

ANALYTICAL PROBLEM

An immunoassay for the determination of the drug morphine involves the use of an anti-morphine antibody which is bound to the plastic walls of the assay tube. Competition occurs between the test morphine and a constant amount of morphine labelled with the enzyme peroxidase for the limited amount of antibody. This results in variable amounts of the labelled morphine being bound by the antibody, and the proportion bound depends on the concentration of the test morphine.

Other related compounds may be present in the test sample, e.g. morphine-6-glucuronide (a metabolite of morphine) and codeine (a proprietary opiate). Thus it is necessary to confirm the specificity of the antibody for morphine and show that the method provides a quantitative basis for its measurement.

The following question must be answered:

- What is the cross-reactivity of the antibody with morphine-6-glucuronide and codeine?

INVESTIGATIONS

Known amounts of morphine standards (100–600 ng ml^{-1}) and the two compounds to be investigated were assayed. This involved adding each sample to a fixed amount of the enzyme labelled morphine in an assay tube and allowing them to react. At the end of the incubation time the tubes were washed and the enzyme activity of the bound fraction was assayed spectrophotometrically. The results were reported as absorbance change per minute.

DATA

Calibration using standard morphine

Standard (ng ml^{-1})	*Reaction rate* (A min^{-1})
0	0.53
	0.54
	0.55
50	0.39
	0.39
	0.38
150	0.24
	0.23
	0.24
300	0.18
	0.16
	0.16

(continued)

600	0.12
	0.12
	0.12

Effect of potential cross-reactants

Cross-reactant (ng ml^{-1})	*Reaction rate* (A min^{-1})
Morphine-6-glucuronide	
100	0.32
	0.32
300	0.20
	0.20
500	0.16
	0.15
1000	0.11
	0.11
Codeine	
100	0.26
	0.25
300	0.17
	0.16
500	0.12
	0.12
1000	0.08
	0.08

SOLVING THE PROBLEM

Only refer to this section when you have made an attempt to answer the question.

1. Construct a calibration curve for the standard morphine, plotting the percentage of labelled morphine bound compared to that bound at zero concentration of morphine (B/B_0) against the concentration of standard morphine.

2. Use this curve to calculate the 'apparent' morphine concentration for each of the three compounds, expressing the result as a percentage of the actual concentration.

PROBLEM 4 SELECTION OF AN APPROPRIATE METHOD ON THE BASIS OF ANALYTICAL PERFORMANCE *(TEST QUESTION)*

INTRODUCTION

The precision, sensitivity and working range of two methods which are available to measure the same analyte, may differ considerably. Before a method can be considered appropriate for a particular application these indicators of performance must be investigated. Refer to 'Assessment of quantitative methods' (p. 1) for background information and also to *Analytical Biochemistry* (§ 1.2.2, 1.2.3).

ANALYTICAL PROBLEM

Two methods (A and B) are under investigation for the measurement of the hormone cortisol. It is expected that sample concentrations will normally be greater than 4 mmol l^{-1} and an appropriate method must be selected.

The following questions need to be answered:

A Which method has the most suitable working range?
B Are the levels of precision and sensitivity acceptable?

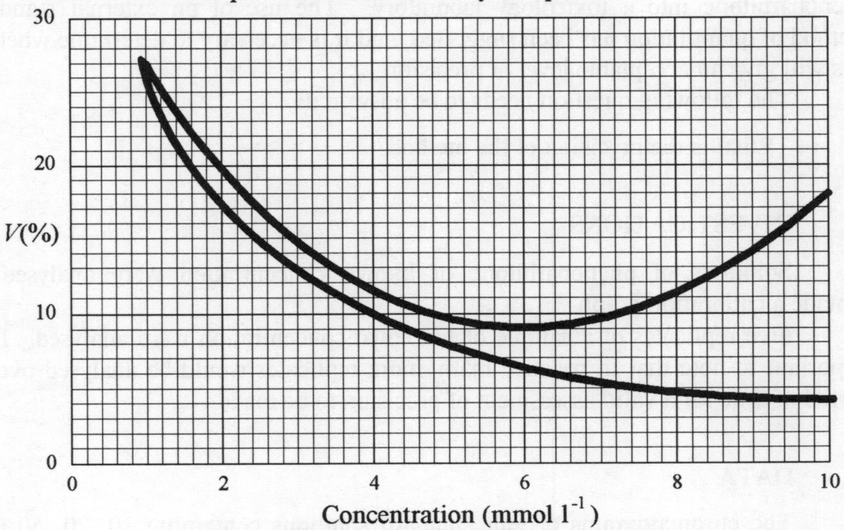

Fig. 1 Concentration vs Coefficient of variation

INVESTIGATIONS

Twenty replicate samples of each of a series of cortisol standards (2–10 mmol l^{-1}) were analysed by the two different methods.

DATA

For each set of replicates the coefficient of variation (*V*) was calculated and plotted against the value of the standard solution.

Figure 1 shows a graph of the results obtained.

PROBLEM 5 ASSESSMENT OF THE PRECISION OF AN HPLC METHOD *(TEST QUESTION)*

INTRODUCTION

The assessment of precision of an analytical method is an essential step in the process of introducing a new method into the laboratory. Quantitation in HPLC involves the use of either an external or an internal standard and the level of precision afforded by these two methods may differ. Refer to 'Assessment of Quantitative Methods' (p. 1) and 'Gas Liquid Chromatography' (p. 26) for background information and also to *Analytical Biochemistry* (§ 1.2.2, 1.2.3).

ANALYTICAL PROBLEM

It is proposed to introduce an HPLC method for the analysis of phenobarbitone into a toxicology laboratory. The use of an external standard method of quantitation has been suggested, and it is necessary to determine whether this will give an acceptable level of precision.

The following question needs to be answered:

■ What is the precision of the method?

INVESTIGATIONS

Solutions of phenobarbitone of known concentration were analysed to provide a calibration graph.

Five replicates of a sample of unknown concentration were analysed. It is important to note that in practice many more replicates would be analysed over a period of time for a valid assessment of precision to be made.

DATA

The chromatograms of four standard solutions containing 10, 20, 50 and 100 mg l^{-1} phenobarbitone are shown in Figure 2a.

The chromatograms of five replicate analyses of an unknown sample are shown in Figure 2b.

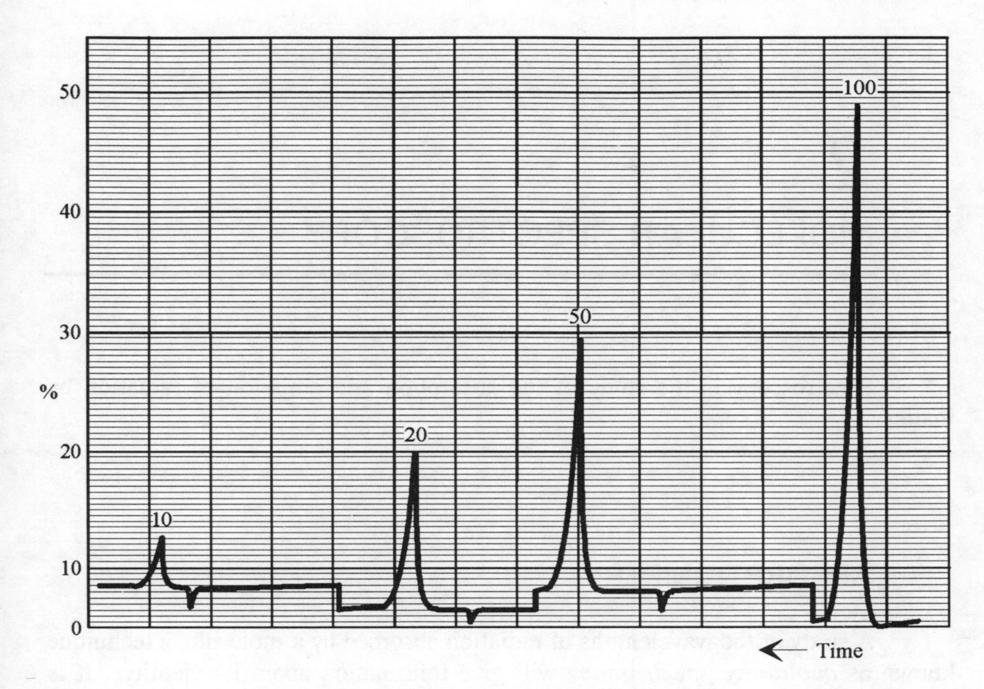

Fig. 2 Chromatograms of (a) standard solution of phenobarbitone. The concentration of each sample in mg l^{-1} is indicated as a number above each peak.

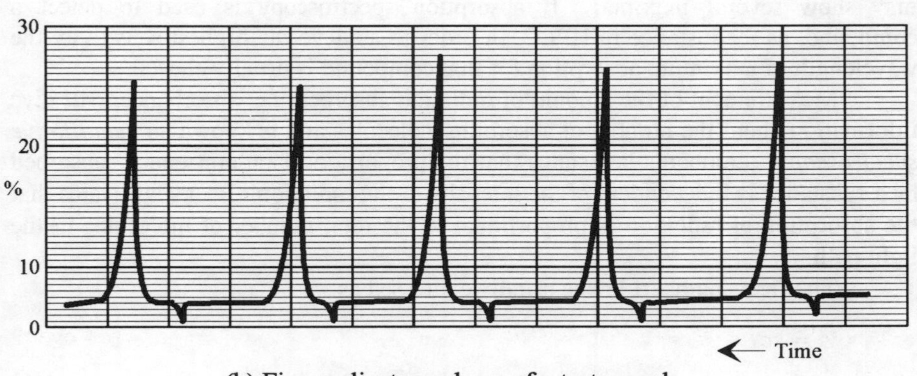

(b) Five replicate analyses of a test sample.

2

MOLECULAR SPECTROSCOPY

Spectroscopy is the study of the absorption and emission of radiation by matter.

ABSORPTIOMETRY

A study of the wavelengths of radiation absorbed by a molecule, a technique known as *qualitative spectroscopy* will give information about its identity; it is most easily represented as a spectrum, i.e. a plot of wavelength against the amount of radiation absorbed. The wavelength showing the greatest absorption, a peak on an absorption spectrum, is known as the absorption maximum. Some compounds may show several maxima. If absorption spectroscopy is used to detect a compound, as for instance in HPLC, in order to achieve the highest sensitivity the wavelength of maximum absorption for that compound is usually chosen.

Measurement of the amount of radiation absorbed, i.e. absorbance, will give information about the number of absorbing molecules and is known as *quantitative spectroscopy*. Lambert's law states that the proportion of radiant energy absorbed by a substance is independent of the intensity of the radiation. Beer's law states that the absorption of radiation is proportional to the total number of molecules in the light path.

The Beer–Lambert law is usually expressed as:

$$A = \varepsilon c l$$

where A is the absorbance,
c is the concentration (mol l^{-1}),
l is the light path (cm)
and ε is the molar absorption coefficient ($l\ mol^{-1} cm^{-1}$).

It is possible to calculate the concentration of a known compound using the Beer–Lambert equation and a previously determined value for ε of that compound from the absorbance of the sample measured at a stated wavelength, which is usually its absorption maximum.

FLUORIMETRY

Molecular fluorescence involves the emission of radiation after excitation caused by the absorption of radiation. The wavelengths of radiation emitted are of a longer wavelength than those absorbed and are useful in the identification of a molecule. Information regarding the emission characteristics of a molecule can be represented by an emission spectrum, i.e. a plot of wavelength against percentage emission. The intensity of the emitted radiation is related to the number of molecules involved but the relationship does not obey the Beer–Lambert equation which is only true for absorption of radiation.

The investigation of the fluorescent properties of a compound must start with a study of its absorption characteristics, and culminate in the determination of the optimum wavelength for producing fluorescence (excitation wavelength) and the wavelength of the radiation which is then emitted (emission wavelength).

PROBLEM 6 AN INVESTIGATION OF THE ABSORPTION AND FLUORESCENT CHARACTERISTICS OF A COMPOUND

INTRODUCTION

Molecular absorption spectrophotometry is frequently used in quantitative methods but fluorimetry potentially offers increased sensitivity for compounds which exhibit fluorescence. In order to develop a method of analysis using either of these techniques, the basic spectroscopic characteristics of the compound must be known. Refer to 'Molecular Spectroscopy' (p. 12) for background information and to *Analytical Biochemistry* (§ 2.2, 2.4).

ANALYTICAL PROBLEM

It has been reported that flavin mononucleotide (FMN) shows fluorescent properties, and it is planned to develop an assay method for this compound using a spectrofluorimeter. In order to be able to use fluorescence as the basis of a quantitative method it is necessary to determine the wavelength of the radiation which causes maximum fluorescence (excitation maximum) and the wavelength of emitted radiation (emission maximum). Alternatively, an absorption method could be used; in which case it would be necessary to determine the molar absorption coefficient of FMN at its absorption maxima to permit quantitation by absorption spectroscopy.

The following questions must be answered:

A What are the absorption maxima of FMN and what is the molar absorption coefficient at each maximum?

B What are the excitation and emission maxima for FMN?

INVESTIGATIONS

A solution of FMN containing 10.0 mg l^{-1} was prepared and its absorption spectrum determined using a spectrophotometer.

The emission by FMN was monitored using a spectrofluorimeter after setting the excitation wavelength to each of the absorption maxima in turn.

The emission wavelength of the spectrofluorimeter was set at the value determined above and the range of exciting radiation was scanned to determined the wavelength inducing the greatest degree of fluorescence.

DATA

The spectra are shown in Figure 3. The RMM for FMN is 478 da.

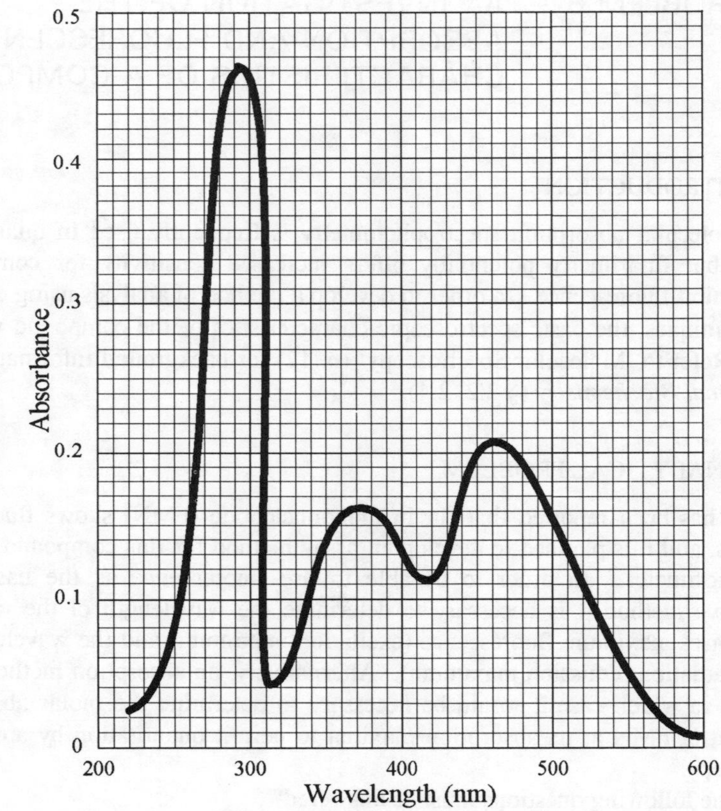

Fig. 3 (a) Absorption spectrum of FMN

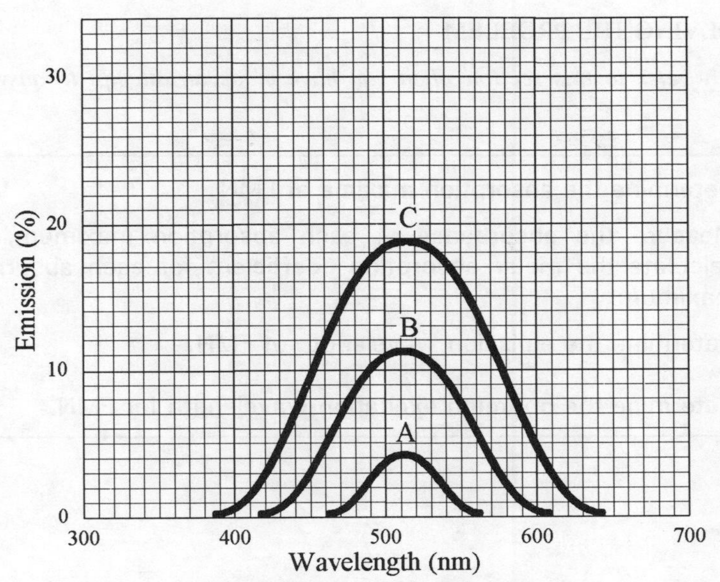

(b) Emission spectra of FMN.
The fluorescence emission of FMN initiated by excitation at three different wavelengths: A 290 nm, B 375 nm, C 460 nm.

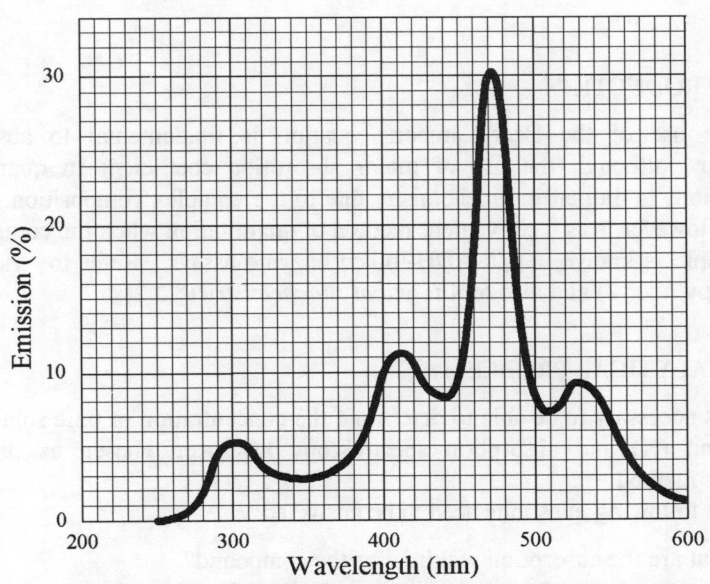

(c) Excitation spectrum of FMN monitored by emission at 515 nm.

SOLVING THE PROBLEM

Only refer to this section when you have made an attempt to answer the question.

1. Determine the absorption maxima of FMN.
2. Measure the absorbance at each absorption maximum, and calculate the molar absorption coefficient for each absorption maximum.
3. Determine the emission wavelength of FMN.
4. Determine the optimum excitation wavelength for FMN.

PROBLEM 7 CALCULATION AND USE OF MOLAR ABSORPTION COEFFICIENT

INTRODUCTION

The use of the Beer–Lambert equation is fundamental to absorption spectroscopy, although the use of molar absorption coefficient in quantitative determinations is limited in biochemistry due to the complex composition of most samples. However, it is a convenient method of quantitation when the composition of the sample is known, e.g. NADH-linked enzyme assays. Refer to 'Molecular Spectroscopy' (p. 12) and also to *Analytical Biochemistry* (§ 2.2).

ANALYTICAL PROBLEM

It is necessary to be able to determine the concentration of pure solutions of a compound (X) and absorption spectroscopy has been chosen as the most appropriate method.

The following questions need to be answered:

A What are the absorption maxima for the compound?

B What is the molar absorption coefficient for the compound at each absorption maximum?

C What is the concentration of the unknown sample of the compound (mol l^{-1})?

INVESTIGATIONS

The absorption spectrum of a pure sample of compound X was determined using a solution containing 8 mg dissolved in 100 ml distilled water.

The absorbance of the unknown solution was measured at 257 nm. All measurements of absorbance were made in a cuvette with a light path of 1 cm.

DATA

- The RMM of compound X is 220 da;
- The absorption spectrum of compound X is shown in Figure 4;
- The unknown solution of compound X had an absorbance of 0.95 at 256 nm.

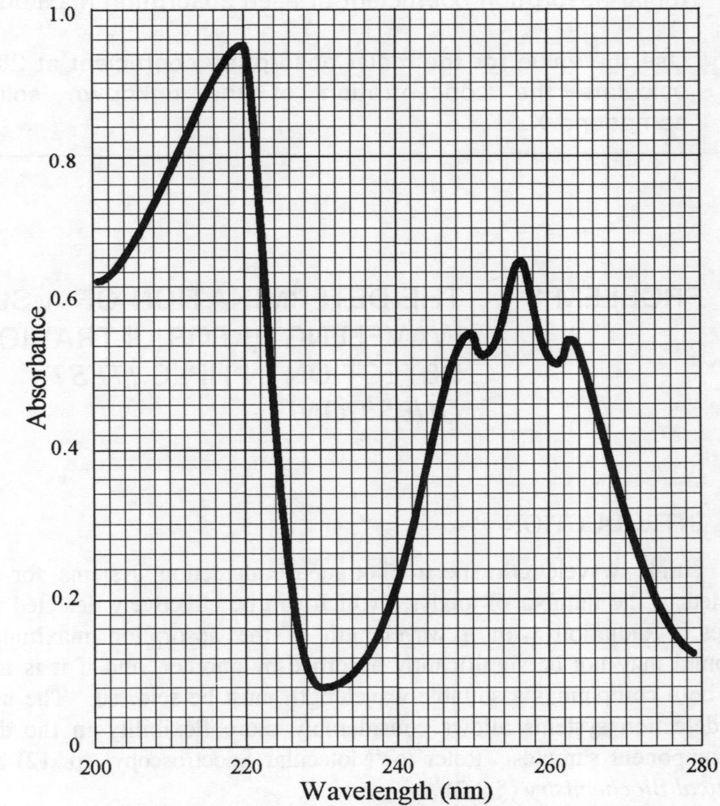

Fig. 4 Absorption spectrum of compound *X*.

SOLVING THE PROBLEM

Only refer to this section when you have made an attempt to answer the question.

1. Examine the absorption spectrum, determine the absorption maxima and read off the absorbance value at each absorption maximum.

2. From the known value for the RMM and the weight dissolved in 1 l, calculate the molar concentration of the solution of compound X used.

3. Using the Beer–Lambert equation, calculate the value for the molar absorption coefficient at each absorption maximum.

4. Use the value for the molar absorption coefficient at 256 nm to calculate the concentration of the unknown solution of compound X.

PROBLEM 8 THE DETERMINATION OF A SUITABLE WAVELENGTH FOR ULTRAVIOLET DETECTION IN HPLC *(TEST QUESTION)*

INTRODUCTION

Single wavelength spectrophotometric detection systems for HPLC are restricted in the number of analytes which can be effectively detected in complex samples. Radiation with a wavelength of the absorption maximum for one compound may not be significantly absorbed by another, and if it is necessary to detect both compounds a suitable wavelength must be selected. The use of diode array detection systems allows considerably more flexibility in the detection of multicomponent samples. Refer to 'Molecular Spectroscopy' (p. 12) and also to *Analytical Biochemistry* (§ 2.2).

ANALYTICAL PROBLEM

It is necessary to determine the concentration of the vitamins riboflavin and niacin in a food. Reverse phase HPLC has been chosen as the method of analysis and a suitable detection wavelength must be selected.

The following questions need to be answered:

A What is the most appropriate wavelength for the simultaneous detection of both vitamins?

B What is the molar absorption coefficient of each compound at the wavelength selected?

INVESTIGATIONS

The absorption spectra for both vitamins were determined.

DATA

The absorption spectra for solutions of riboflavin and niacin are shown in Figure 5 (a and b).

Chromatogram	*Analyte*	*Concentration* ($mg\ l^{-1}$)	*RMM* (da)
A	Niacin	12.0	123.1
B	Riboflavin	3.0	376.4

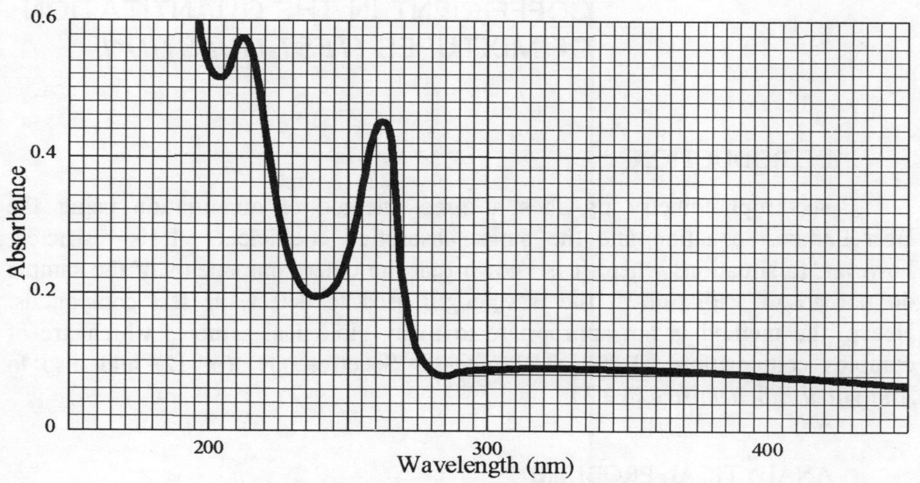

Fig. 5 Absorption spectrum of (a) niacin

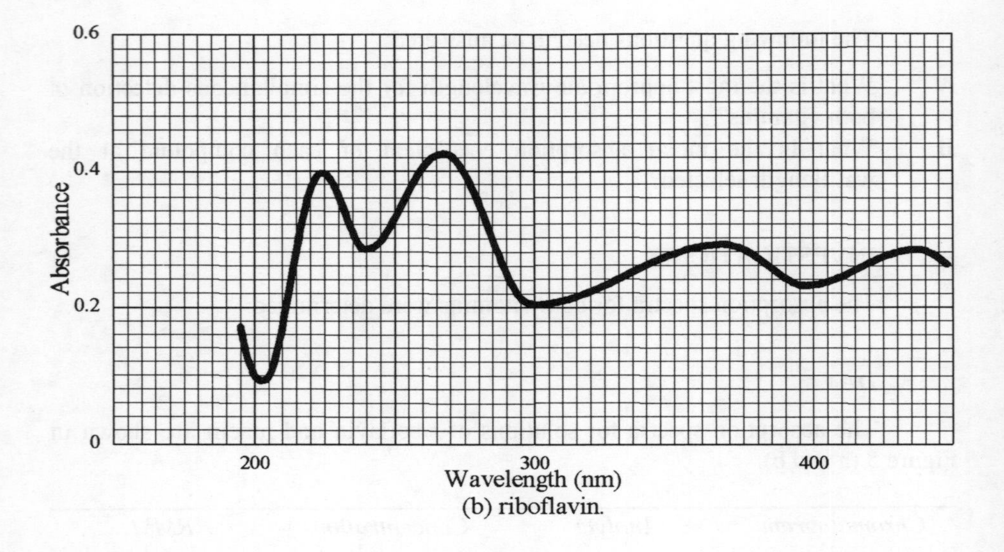

(b) riboflavin.

PROBLEM 9 THE USE OF MOLAR ABSORPTION COEFFICIENT IN THE QUANTITATION OF MIXTURES *(TEST QUESTION)*

INTRODUCTION

Spectrophotometry provides a direct method of quantitation using the Beer–Lambert equation and the molar absoıption coefficient. It is restricted, however, to situations where it is known that the other components of the sample do not absorb radiation at the wavelength used to detect the test compounds. Hence, the method is not appropriate to many biological samples which are of complex composition. Refer to 'Molecular Spectroscopy' (p. 12) and also to *Analytical Biochemistry* (§ 2.3).

ANALYTICAL PROBLEM

It is necessary to determine the actual composition of a solution containing a mixture of nicotinamide-adenine dinucleotide (reduced) (NADH) and cytidine monophosphate (CMP). Absorption spectroscopy has been chosen as an appropriate method.

The following questions need to be answered:

A What is the molar absorption coefficient for each compound at its absorption maximum?

B What is the concentration of each compound in the mixture (mol l^{-1})?

INVESTIGATIONS

Solutions of each of the two compounds were prepared from the pure solid form and their absorption spectra recorded.

The absorbance of the sample was measured at 340 nm and 260 nm.

DATA

Figure 6 shows the absorption spectra of a solution of NADH containing 0.046 mol l^{-1} and a solution of CMP containing 0.05 mol l^{-1}.

The means of three absorbance measurements of the unknown sample at the two wavelengths specified are given below.

Wavelength (nm)	*Absorbance*
340	0.37
260	1.50

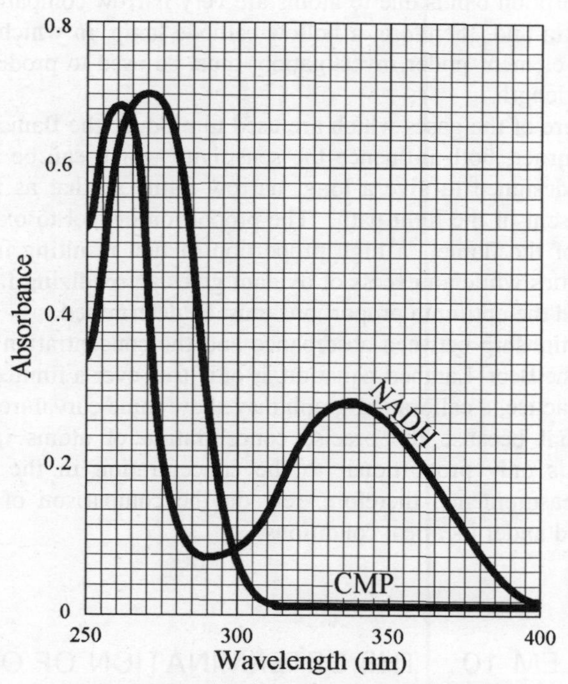

Fig. 6 Absorption spectra of nicotinamide-adenine dinucleotide (reduced) (NADH) and cytidine monophosphate (CMP).

3 ATOMIC SPECTROSCOPY

When atoms are dispersed in a flame, the vast majority remain in the ground, or most stable, state and only a very small proportion are thermally excited. If a beam of light is passed through the flame, atoms in the ground state will absorb the appropriate wavelength. The proportion of radiation which is absorbed by the atoms in the flame is related to the number of atoms present in a manner directly comparable to molecular absorption spectrophotometry.

The absorption bands due to atoms are very narrow compared to molecular absorption spectra and, therefore, a hollow cathode lamp, in which the cathode is coated with the element under investigation, must be used to produce radiation of the correct wavelength.

The nature of the gases which are used to produce the flame and the actual design of the burner, both influence the sensitivity which can be achieved. The burner head is designed to give a long, narrow flame so that as many atoms as possible are present in the light path. The proportion of fuel to oxidant alters the characteristics of the flame; a high proportion of fuel resulting in a flame with reducing properties while an excess of oxidant gives an oxidizing flame. For each element analysed the optimum proportions must be determined.

The relationship between absorbance and the concentration of the sample, as specified by the Beer–Lambert equation, is only true over a limited concentration range and, in practice, a calibration graph may show some curvature. A calibration graph is essential because the precise concentration of atoms in the flame is unknown, and is only proportional to the concentration in the liquid sample. Quantitative measurements, therefore, rely on the comparison of standards and samples analysed under identical conditions.

PROBLEM 10 THE DETERMINATION OF OPTIMAL INSTRUMENT SETTINGS FOR ATOMIC ABSORPTION SPECTROSCOPY

INTRODUCTION

Atomic absorption spectroscopy is a useful technique for the determination of many cations of biological significance. Optimal assay conditions differ appreciably for different elements and it is essential that all are investigated during the development of a method. Other substances present in the sample may affect the results obtained and steps should be taken to minimise these interference effects. Refer to 'Atomic Spectroscopy' (p. 22) for background information and to *Analytical Biochemistry* (§ 2.5.3).

ANALYTICAL PROBLEM

It is intended to measure calcium in serum samples using atomic absorption spectroscopy. Calcium absorbs radiation at 422.7 nm and is particularly amenable to quantitation by atomic absorption spectroscopy. The optimal ratio of fuel to oxidant in the flame for this element must be determined, and any potential interference effects investigated, before the method can be used routinely.

The following questions need to be answered:

A What ratio of fuel (acetylene) to oxidant (air) gives maximum sensitivity?

B Does the presence of common anions in the sample affect the results obtained?

C How can any such interference be reduced?

INVESTIGATIONS

With the monochromator of the atomic absorption spectrophotometer set at 422.7 nm, a solution of calcium carbonate (0.05 mmol l^{-1}) was aspirated into the flame and absorbance readings taken at different fuel to oxidant ratios.

Using the selected fuel to oxidant ratio a range of mixtures, which had been prepared to investigate potential interference effects, were aspirated and the absorbance values recorded.

DATA

Gas flow rate (ml min^{-1})		*Absorbance* (422.7 nm)
Air (oxidant)	Acetylene (fuel)	Calcium 0.05 mmol l^{-1}
20	5	0.9
20	10	1.1
20	15	1.0
20	20	0.8
20	25	0.6
20	30	0.5

Component	Tube number						
	1	2	3	4	5	6	7
	Volume added (ml)						
Distilled water	2	1	1	1	–	–	6
Calcium carbonate ($0.05\ mmol\ l^{-1}$)	5	5	5	5	5	5	–
Sodium chloride ($5\ mmol\ l^{-1}$)	–	–	1	–	1	–	–
Sodium phosphate ($5\ mmol\ l^{-1}$)	–	–	–	1	–	1	–
Lanthanum chloride ($20g\ l^{-1}$)	–	1	–	–	1	1	1
Absorbance	0.8	0.8	0.8	0.3	0.8	0.8	0

SOLVING THE PROBLEM

Only refer to this section when you have made an attempt to answer the question.

1. Plot a graph of absorbance against fuel flow rate and determine the optimum value.
2. Determine the effect of phosphate and chloride on the measurement of calcium.
3. Determine the effect of lanthanum.

PROBLEM 11 QUANTITATION BY ATOMIC ABSORPTION SPECTROSCOPY *(TEST QUESTION)*

INTRODUCTION

Atomic absorption spectroscopy is useful for the determination of many elements in relatively crude samples. Tissues can be homogenized and, after centrifugation to remove debris, the resulting solution can, sometimes, be analysed directly. However, it is often preferable to remove the protein component prior to analysis. Refer to 'Atomic Spectroscopy' (p. 22) and also *Analytical Biochemistry* (§ 2.5.3, 11.3.1).

ANALYTICAL PROBLEM

It is necessary to determine the calcium content of a specimen of plant tissue. Atomic absorption spectroscopy has been used with optimal instrumental settings and the relevant absorbance measurements have been taken.

The following questions need to be answered:

A What is the calcium content of the tissue extract (mmol l^{-1})?
B What was the calcium content of the original tissue (mmol g^{-1})?

INVESTIGATIONS

10 g of tissue was homogenized in a calcium-free buffer. The total volume was 50 ml, and the homogenate was centrifuged to remove debris. Ten ml of the clear supernatant fluid was mixed with 10 ml of trichloroacetic acid (100 g l^{-1}) to precipitate the protein, filtered and the clear filtrate used for analysis.

A series of standard calcium solutions were prepared, these and the sample were treated as follows:

- 10 ml standard calcium or sample;
- 1 ml lanthanum chloride solution (20 g l^{-1}).

All solutions were analysed in duplicate using an atomic absorption spectrophotometer at 422.7 nm. The instrument was zeroed on a solution containing 10 ml distilled water instead of the sample.

DATA

Solution	*Absorbance*	
Standard calcium (mmol l^{-1})		
0.1	0.22	0.24
0.2	0.46	0.49
0.3	0.66	0.67
0.4	0.80	0.79
0.5	0.90	0.91
Tissue extract	0.50	0.51

4

GAS LIQUID CHROMATOGRAPHY

Gas liquid chromatography (GLC) depends upon the partition of a solute between two physical states or phases, i.e. gaseous and liquid. In GLC the stationary liquid phase is coated on an inert, particulate supporting medium which is contained in stainless-steel or glass columns. The test compounds are injected into the column, vaporised and blown down the column by the carrier gas, most frequently nitrogen, argon or helium. Upon initial contact of each analyte with the stationary phase, an equilibrium is rapidly established between the amount of analyte which dissolves in the stationary liquid phase and the amount remaining in the vapour. The precise equilibrium position is a characteristic of the analyte and the nature of both the stationary phase and the carrier gas. When the position lies towards solution, the compounds will tend to be retained on the column longer than those whose equilibrium position tends towards the vapour phase. If the temperature of the column is raised the equilibrium position will be displaced towards the vapour phase, with the consequence that the time on the column (retention time) will be shortened.

The sample vapours are detected as they leave the column and the detector response is displayed on a recorder as a trace known as a chromatogram. Individual components have characteristic retention times, i.e. the time taken (or the distance on the chromatogram) from the point of injection to the midpoint of the peak. However, the reproducibility of retention times is significantly affected by chromatographic conditions, e.g. alterations in the gas flow, column temperature, etc. It is, therefore, not adequate to rely on retention times alone for identification purposes.

METHOD DESIGN

The major consideration in selecting a stationary (liquid) phase is its polarity relative to the test compounds. Polarity refers to the molecular characteristic produced by an unequal distribution of electrons (and hence charge) in a molecule. A non-polar stationary phase will tend to retain non-polar solutes while a polar phase will show greater affinity for polar solutes. In non-polar systems, because hydrogen bonding does not occur, elution of the solutes is usually in order of their boiling points. Hydrogen bonding will occur to varying degrees with polar systems and will appreciably modify the elution order because molecules involved in hydrogen bonding will be retarded.

When samples contain a mixture of compounds with a wide range of volatilities, separation may often be more effectively achieved using a temperature gradient rather than a constant temperature (*isothermal*). In *temperature gradient* separation, the analysis is initiated at the lower temperature which is then raised at a specified rate.

Many substances are not initially appropriate for GLC because of their relatively high boiling-points. In such cases it is sometimes possible to chemically modify them to reduce their polarity and lower their boiling-points. This procedure is called *derivatization* and several methods are available.

DETECTION SYSTEMS

A range of detectors is available with varying specificities and sensitivities. For instance, the flame ionization detector (FID) is capable of detecting virtually all organic compounds and shows a lower limit of sensitivity of approximately 1×10^{-9} mol. Electron capture detectors (ECD), however, show greater specificity and sensitivity towards electrophilic compounds.

ANALYTICAL ASPECTS

Although the response of the detector is usually proportional to the concentration of the test substance, it is essential that this is checked using a series of standard solutions. The relationship between the concentration of the analyte and the peak on the chromatogram is, strictly speaking, only valid for peak area measurements. However, it is often more convenient to use peak heights, and this is usually acceptable when all peaks have similar widths.

For *external standardization*, replicate standards of known concentrations of the pure substance under investigation are injected, and the areas of the resulting peaks measured. The sample is then injected and the peak area compared with those of the standards, often by means of a calibration graph. Modern instruments normally perform this operation automatically. The precision of this external standardization method is often very poor, particularly as a result of variation in injection volume or detector response, and internal standardization is necessary.

Internal standardization involves the addition to the sample of a known amount of a reference substance which was not originally present in the sample. This will results in the appearance of an additional peak on the chromatogram.

A detector may respond differently to the test substance and the internal standard. It is therefore necessary to determine the *response factor* (R) of the detector for each test substance relative to the internal standard. This can be done by injecting solutions which contain known amounts of the test substance and the internal standard.

$$\frac{\text{Test peak height}}{\text{Standard peak height}} \times R = \frac{\text{test concentration}}{\text{standard concentration}}$$

For quantitation of the sample, a known amount of the internal standard is introduced into the sample and an aliquot of the mixture is injected. Knowing the concentration of the internal standard (C_s) and the response factor, the test concentration, (C_t) can be calculated:

$$C_t = \frac{\text{test peak height}}{\text{standard peak height}} \times C_s \times R$$

The use of a single response factor assumes that it is constant over a wide concentration range, and it is often more acceptable to determine the response factor for a range of test concentrations. This can be done by incorporating a fixed amount of the internal standard in samples which contain known concentrations of the test substance. For each concentration the ratio of peak heights of the analyte to the internal standard is determined and plotted against concentration. This calibration graph can then be used to determine the amount of test substance in an unknown sample to which the same fixed amount of internal standard has been added.

ASSESSMENT OF COLUMN EFFICIENCY

The ability of a column chromatographic process to separate or resolve two similar compounds is indicated by the *resolution index* (R_s). The extent of the resolution of two separate peaks on a chromatogram may be quantified using the following equation:

$$\text{Resolution } (R_s) = \frac{\text{twice the distance between the two peaks}}{\text{sum of the base width of the two peaks}}$$

The greater the value of R_s, the better the resolution of the two compounds. However, high values are undesirable due to the long analysis times involved.

The efficiency of a column separation process is quoted in terms of the number of *theoretical plates* (N). A theoretical plate is a notional concept which relates to the number of equilibria which have taken place during the separation process. The greater the value for N, the more efficient is the column. The number of theoretical plates for a separation can be calculated as follows:

$$N = \left(\frac{\text{retention distance}}{\text{peak width at half peak height}} \right)^2 \times 5.54$$

The efficiency of separation is also related to column length and the parameter of the *height equivalent to a theoretical plate* (HETP) takes account of this. HETP may be calculated from the value for N and the length of the column as follows:

$$\text{HETP} = \frac{\text{length of the column}}{N}$$

Analytical conditions should be determined which give the greatest separation efficiency. This includes using an appropriate carrier gas flow rate which gives the lowest value for HETP.

PROBLEM 12 THE DETERMINATION OF OPTIMAL GAS FLOW IN GLC

INTRODUCTION

In GLC the separation process depends on the partitioning of the analytes between the liquid and the gas phases. This equilibrium is being continually upset and re-established as the carrier gas flows through the column. The flow rate of the carrier gas, therefore, is an important factor in the efficiency of the separation. If the gas flow is too great, effective equilibration will not be achieved and poor separation will result. For maximum efficiency it is necessary to perform separations at the optimal gas flow rate for the analysis. Refer to 'Gas Liquid Chromatography' (p. 26) for background information and to *Analytical Biochemistry* (§ 3.2.2, 3.2.3 and 12.7.4).

ANALYTICAL PROBLEM

Fatty acids can be separated as their methyl ester derivatives by GLC at 190°C using a PEG (polyethylene glycol) column and flame ionization detection. In setting up such a method for the analysis of fatty acids it is necessary to determine the optimal gas flow rate.

The following question must be answered:

- What flow rate gives a minimum value for HETP (height equivalent to a theoretical plate)?

INVESTIGATIONS

The methyl ester of palmitic acid was injected onto a 10% PEG column 500 cm long at 190°C using different flow rates of the nitrogen carrier gas.

DATA

The chromatograms are shown in Figure 7 (a-f).

Chromatogram	*Gas flow* (ml min^{-1})
A	4.0
B	3.5
C	3.0
D	2.5
E	2.0
F	1.5

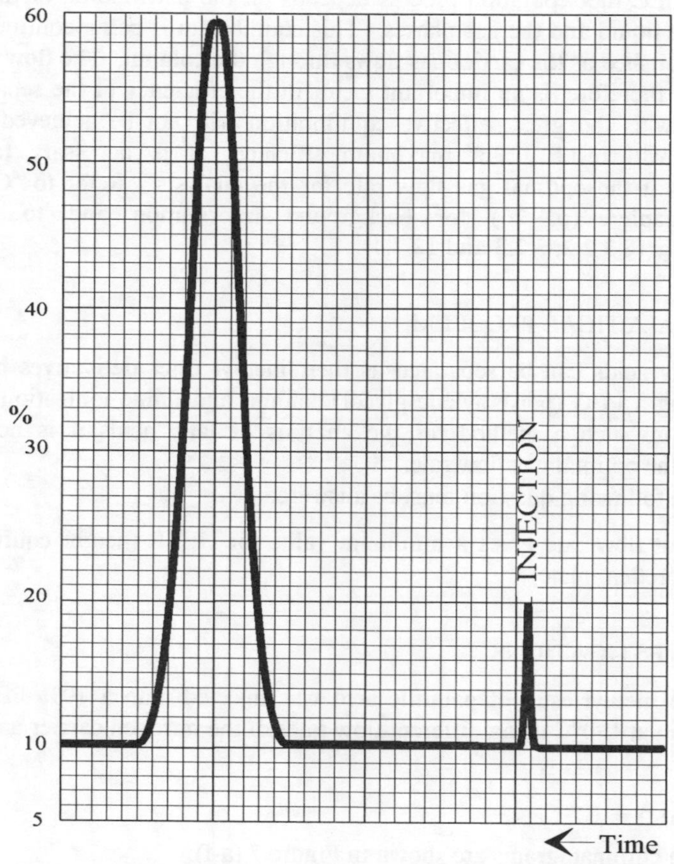

Fig. 7 (a) Chromatogram A-gas flow rate 4.0 ml min^{-1}

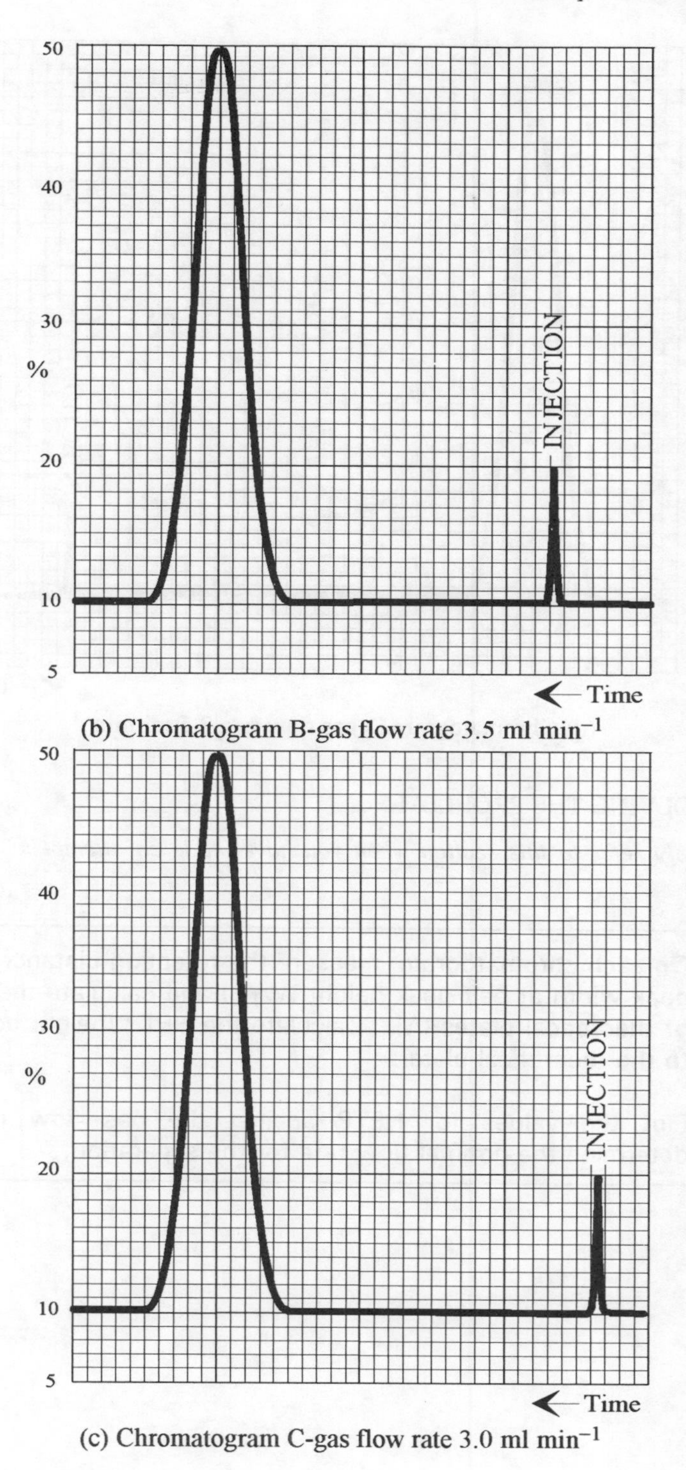

(b) Chromatogram B-gas flow rate 3.5 ml min^{-1}

(c) Chromatogram C-gas flow rate 3.0 ml min^{-1}

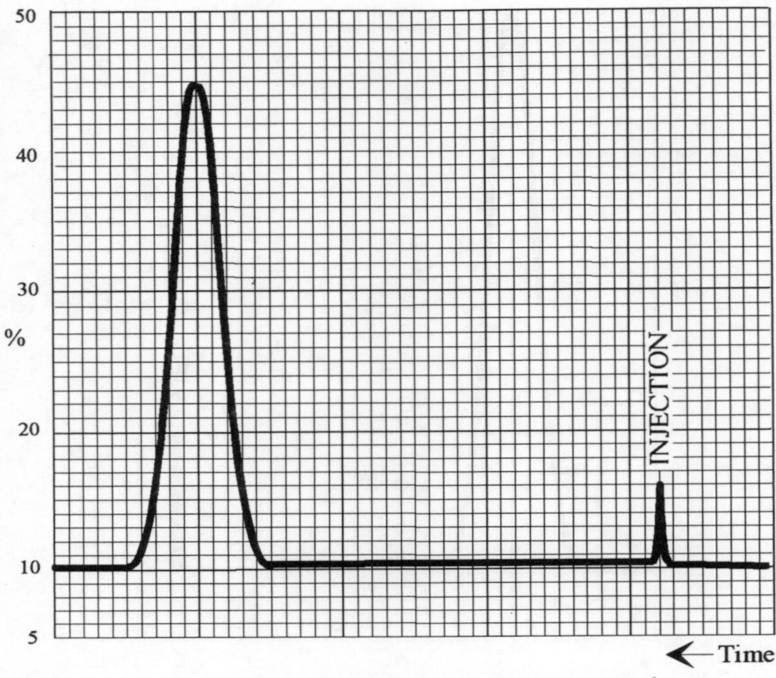

(d) Chromatogram D-gas flow rate 2.5 ml min^{-1}

SOLVING THE PROBLEM

Only refer to this section when you have made an attempt to answer the question.

1. On each chromatogram, measure the retention distance and the peak width at half peak height ($W_{1/2}$) and calculate the number of theoretical plates (N). Calculate the HETP (height equivalent to the theoretical plate).

2. Plot the values for HETP against the gas flow rate and determine the optimal flow rate for the separation.

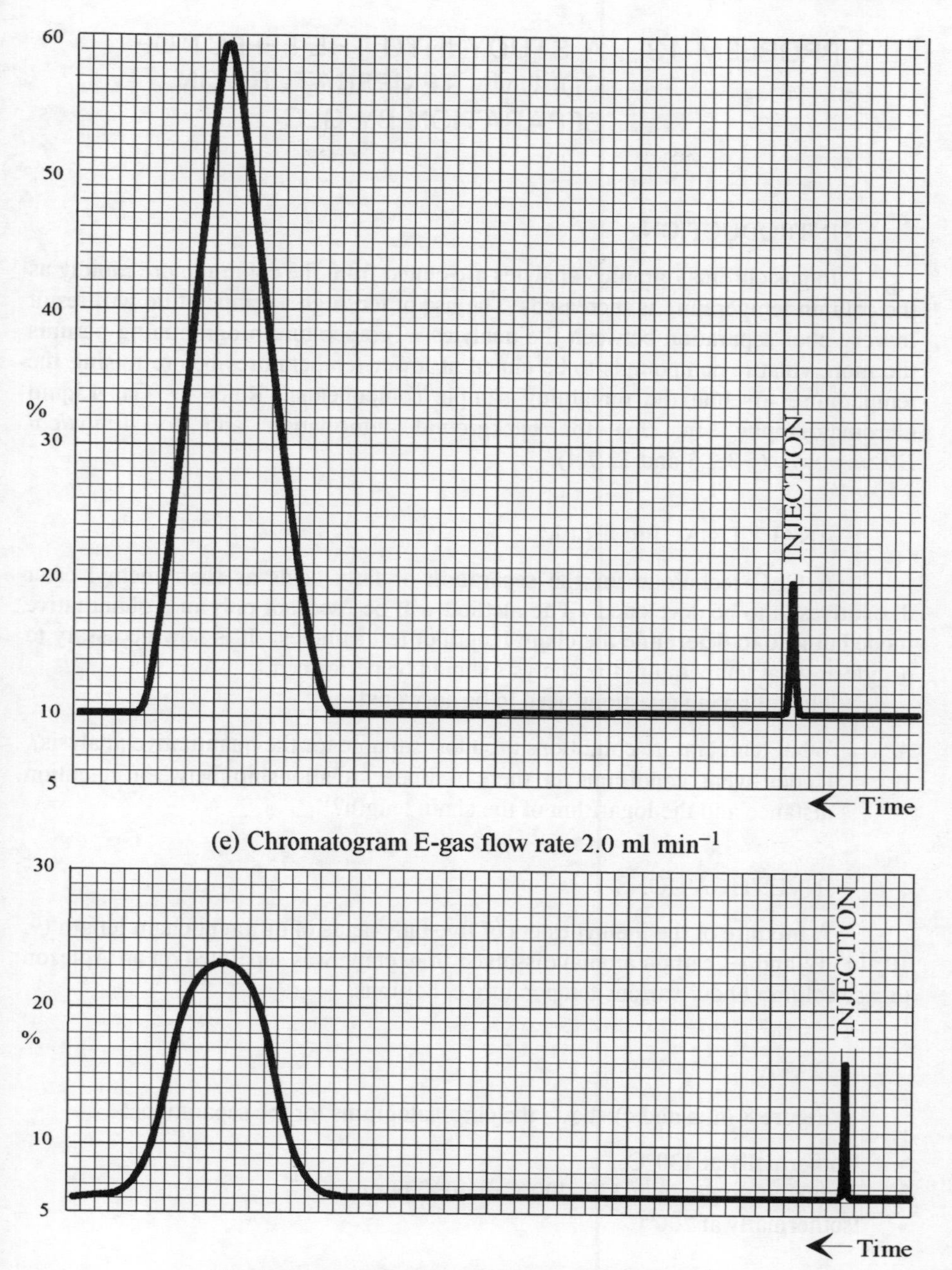

(e) Chromatogram E-gas flow rate 2.0 ml min^{-1}

(f) Chromatogram F-gas flow rate 1.5 ml min^{-1}

PROBLEM 13 A STUDY INTO THE EFFECT OF COLUMN TEMPERATURE ON SEPARATION BY GLC

INTRODUCTION

In gas chromatography, analytes are eluted from the column more rapidly as the column temperature is increased. The use of too high a temperature will result in very poor separation between the analytes. Temperature programming permits the more volatile compounds to be eluted at lower temperatures before raising the temperature to elute the remaining sample components. Refer to 'Gas Liquid Chromatography' (p. 26) for background information and to *Analytical Biochemistry* (§ 3.2.3 and 12.7.4).

ANALYTICAL PROBLEM

A GLC technique for the separation of fatty acids as their methyl ester derivatives is to be developed. The method will be used not only in a quantitative mode but also to determine the chain length of test samples. It is now necessary to decide at what temperature the analysis should be performed.

The following questions need to be answered:

A What temperature conditions are most suitable for the quantitative analysis?

B Under what conditions is there a linear relationship between retention distance and the logarithm of the chain length?

INVESTIGATIONS

A mixture of the methyl esters of five fatty acids of different chain length (9, 10, 11, 12 and 13 carbon atoms) dissolved in heptane was separated on an Apiezon grease column under various temperature conditions.

DATA

Figure 8 (a, b and c) shows the chromatograms for the separations:

- Isothermally at 130°C;
- Temperature gradient 130–200°C at 10°C per min;
- Isothermally at 200°C.

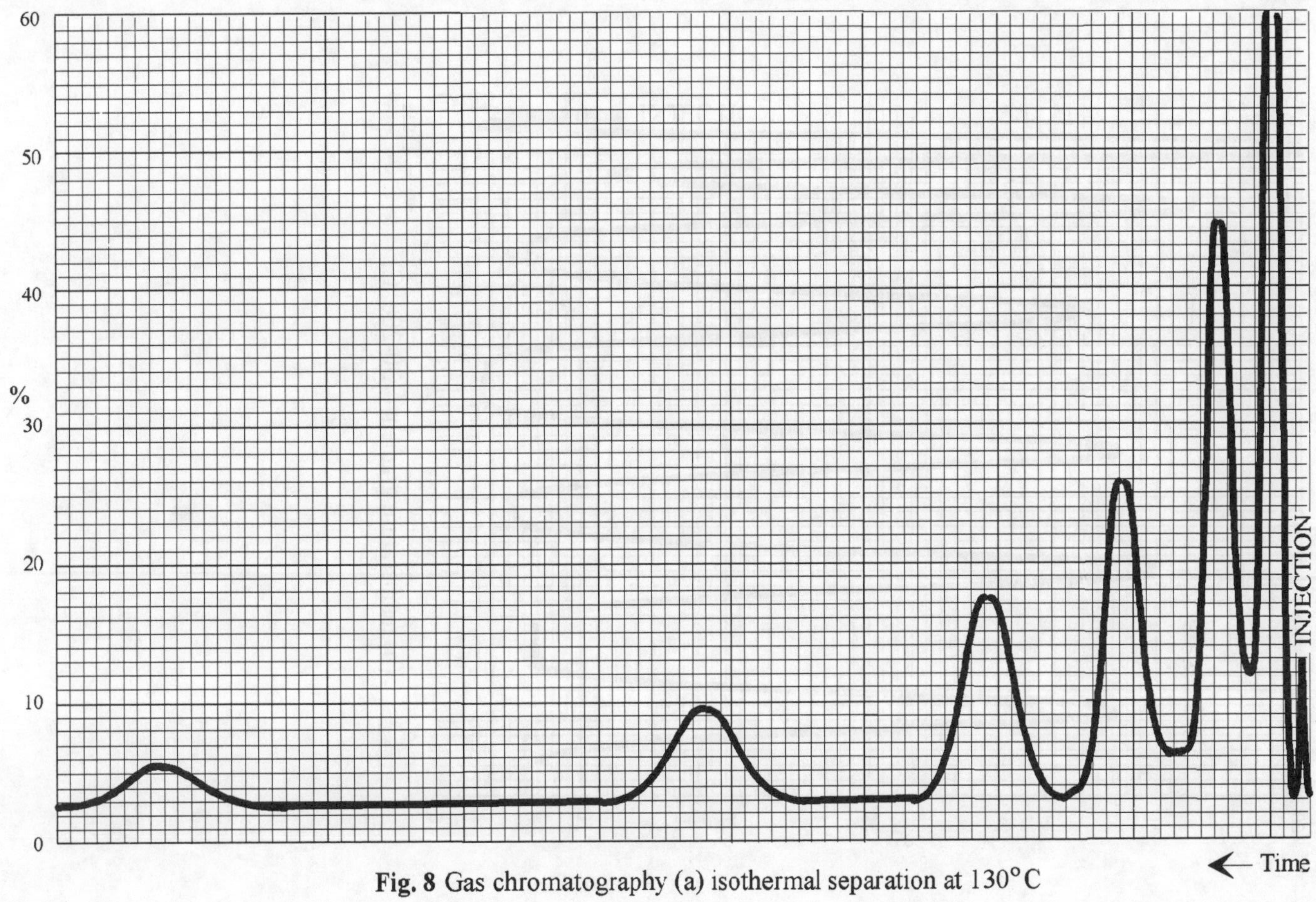

Fig. 8 Gas chromatography (a) isothermal separation at 130°C

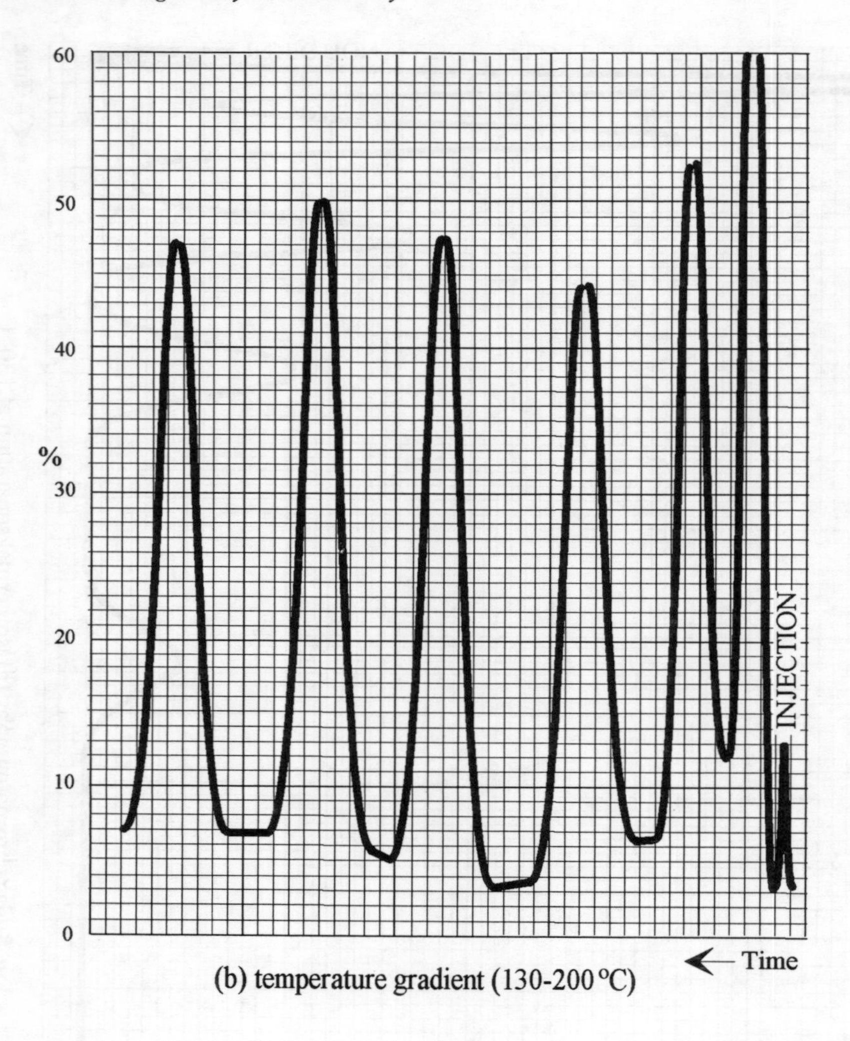

(b) temperature gradient (130-200 °C)

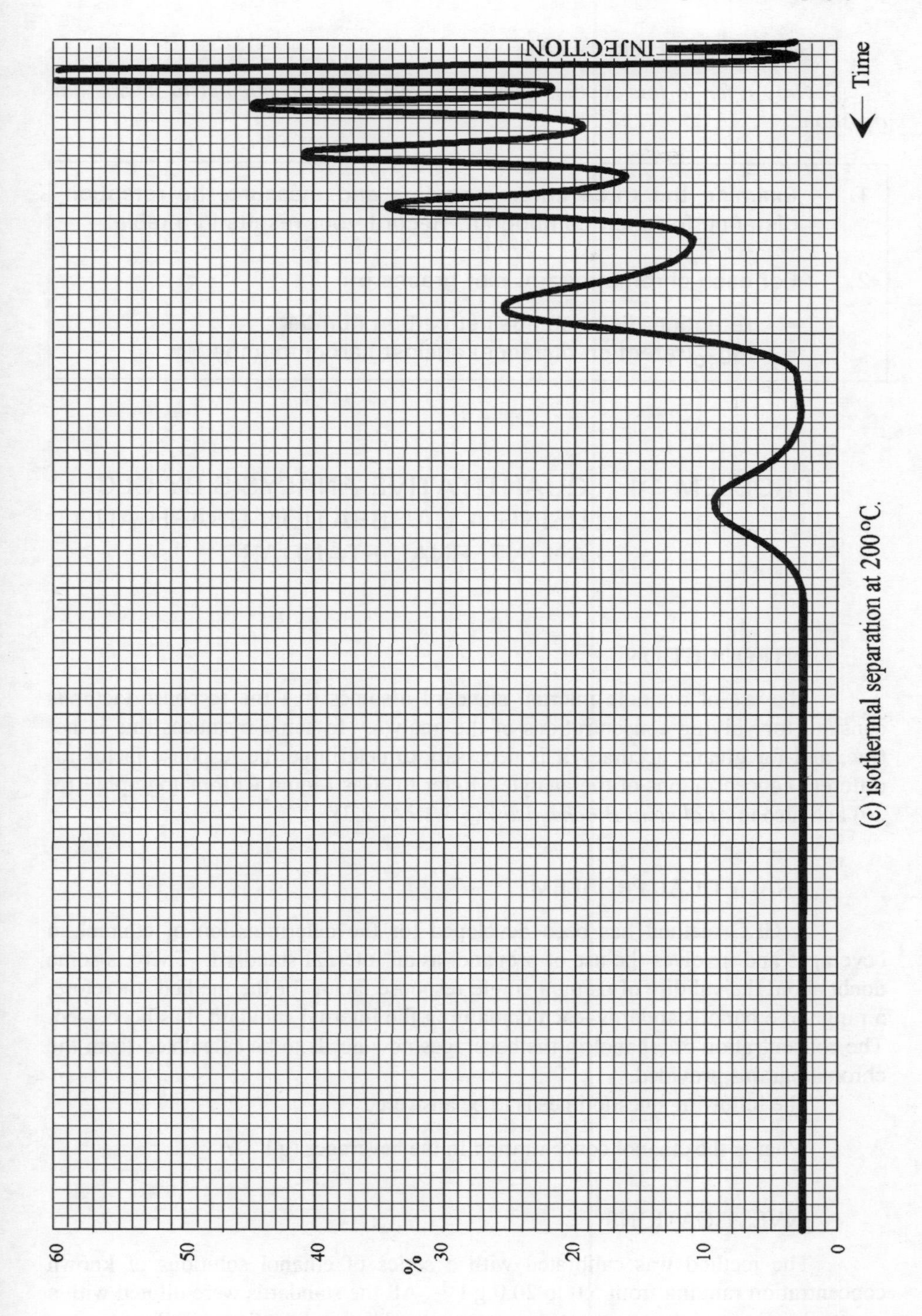

(c) isothermal separation at 200 °C.

SOLVING THE PROBLEM

Only refer to this section when you have made an attempt to answer the question.

1. Examine the three chromatograms and measure the retention distance for each compound. Record your results in a table.

2. For each chromatogram, plot graphs of:
   - retention distance against carbon number;
   - $\log_{10}$(retention distance) against carbon number.

PROBLEM 14 QUANTITATIVE ANALYSIS BY GLC USING A CALIBRATION GRAPH AND AN INTERNAL STANDARD

INTRODUCTION

The use of a single internal standard assumes that the response factor is constant for varying concentrations of the analyte. In some instances this is not true, and for greater accuracy it is necessary to determine the response factor for different concentrations of the analyte. Refer to 'Gas Liquid Chromatography' (p. 26) and also to *Analytical Biochemistry* (§ 3.2.2, 3.2.3).

ANALYTICAL PROBLEM

A GLC method has been developed for the determination of ethanol in beverages and involves the use of propanol as an internal standard. There is some doubt about the validity of using a single response factor for the method, therefore, a range of ethanol standards, each containing the internal standard, has been used. The concentration of ethanol in the beverages now needs to be calculated from the chromatograms provided.

The following question needs to be answered:

A What is the ethanol concentration in the beverages (g l^{-1})?

INVESTIGATIONS

The method was calibrated with a series of ethanol solutions of known concentration ranging from 5.0 to 20.0 g l^{-1}. All the standards were diluted with a constant amount of the propanol internal standard before injection as follows:

- 1.0 ml propanol solution;
- 0.1 ml ethanol standard solution;
- 5.0 μl of the mixture was injected.

Prior to analysis the beverages were diluted with water, as indicated, and 0.1 ml of the each diluted sample was mixed with 1.0 ml of the same propanol internal standard solution prior to injection.

DATA

Sample	*Volume of sample* (ml)	*Volume of distilled water* (ml)
(A) Wine	0.1	0.5
(B) Beer	0.1	0.2
(C) Lager	0.1	0.2
(D) Whisky	0.1	7.0
(E) Vodka	0.1	7.0

The chromatograms for the analyses are shown in Figure 9 (a-c).

SOLVING THE PROBLEM

Only refer to this section when you have made an attempt to answer the question.

1. Measure the peak height of both the ethanol and propanol peaks in the analysis of the ethanol standards.

2. Calculate the ethanol/propanol (E/P) ratio for each standard solution of ethanol and plot the values against the concentration of ethanol.

3. For each beverage sample, calculate the E/P ratio and read off the ethanol concentration from the calibration graph. Correct the answer for the dilution factor introduced before the analysis.

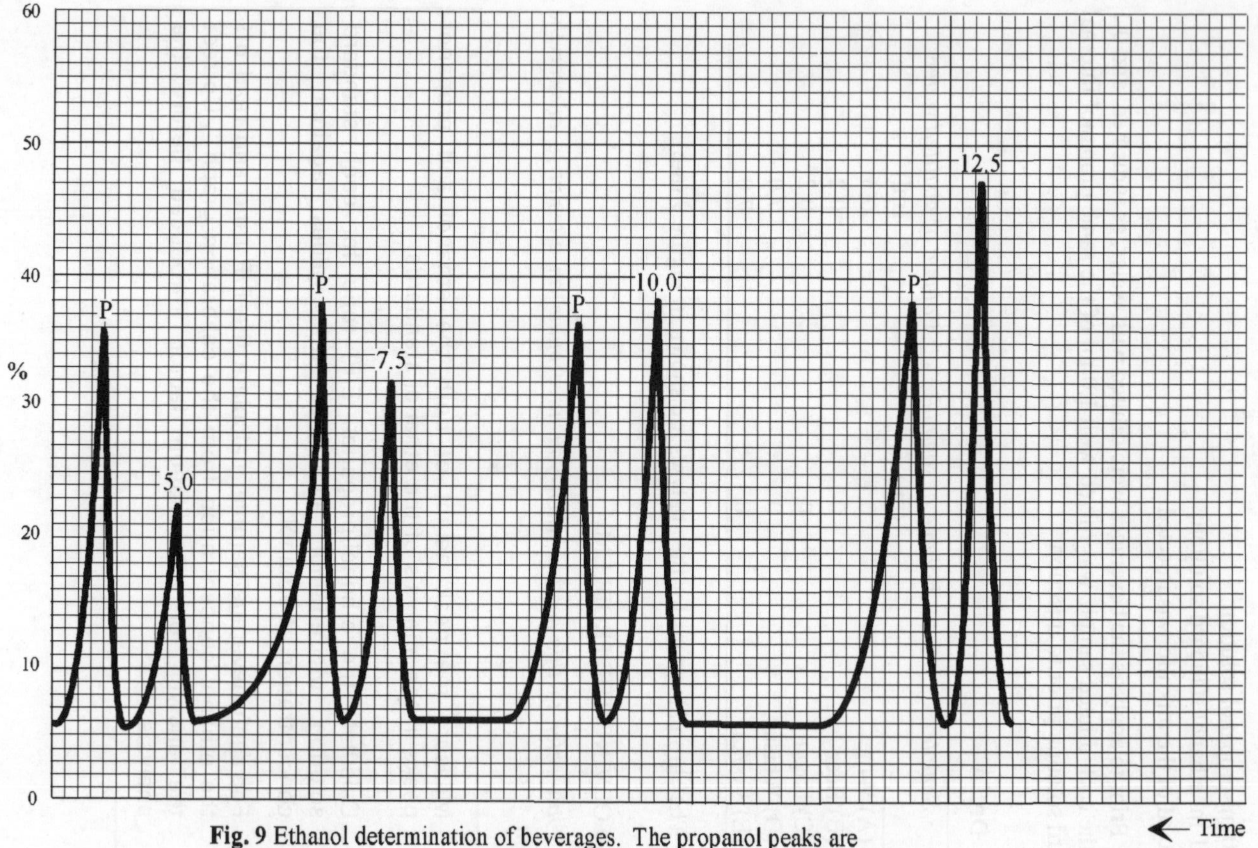

Fig. 9 Ethanol determination of beverages. The propanol peaks are identified by a P, and the ethanol standards by a number which also specifies the ethanol concentration in g l^{-1}. (a) standard solutions of ethanol

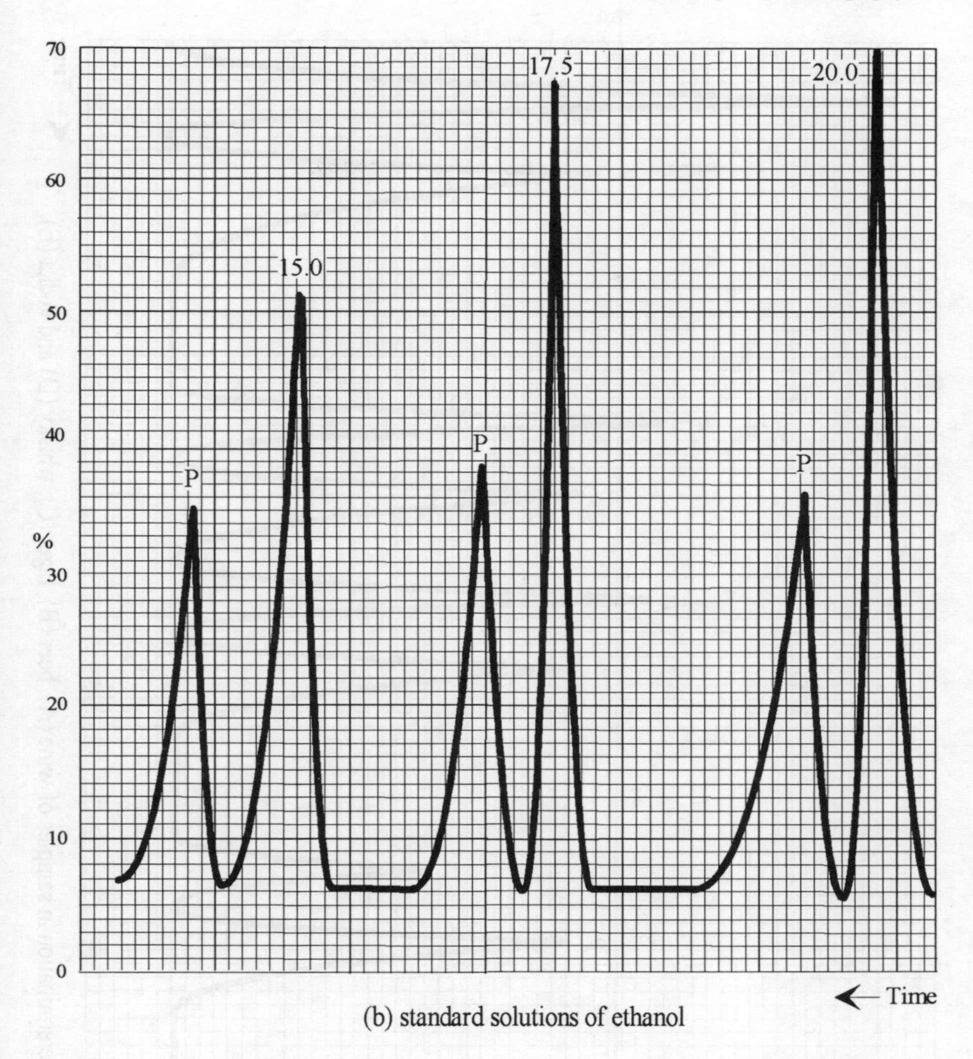

(b) standard solutions of ethanol

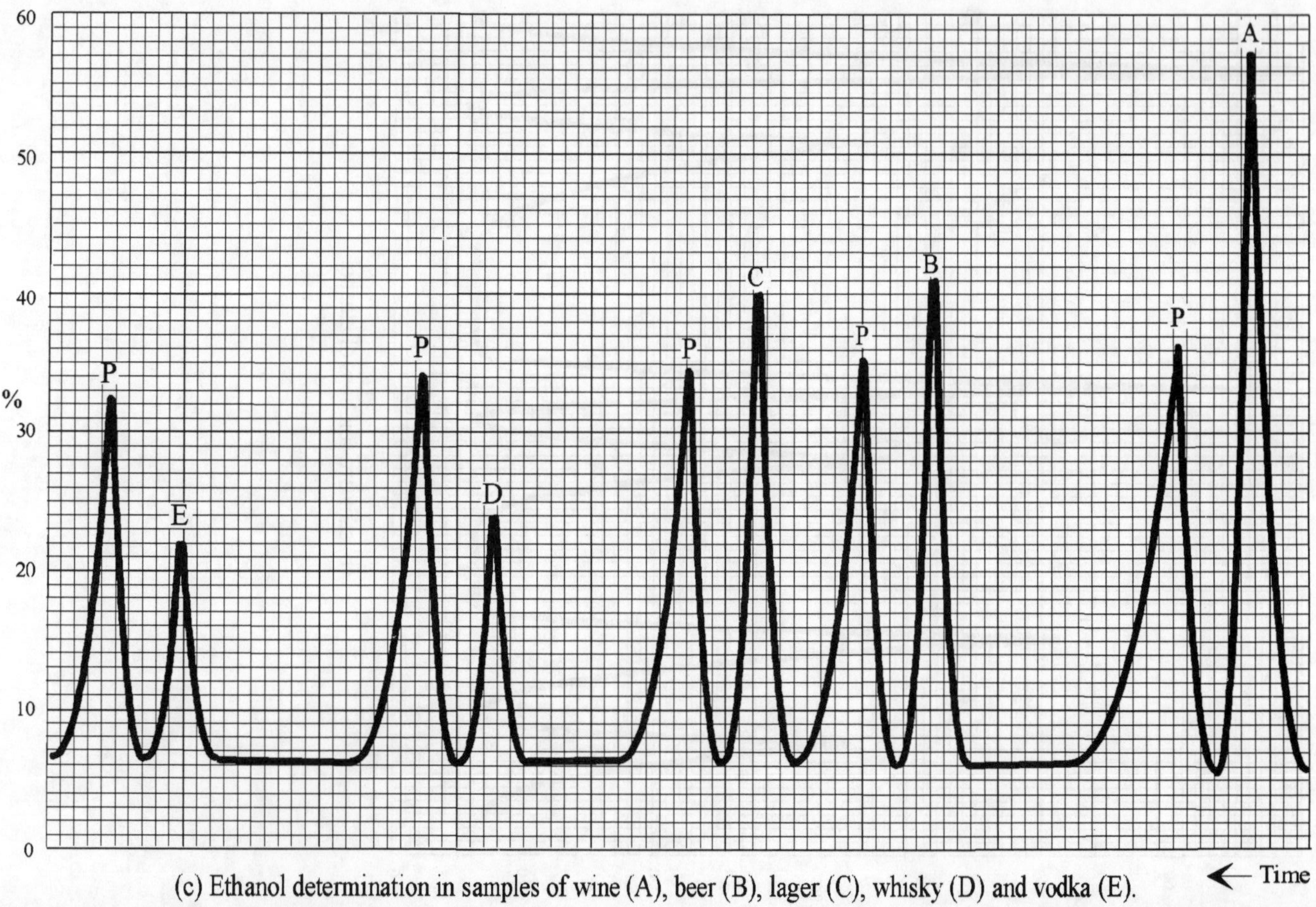

(c) Ethanol determination in samples of wine (A), beer (B), lager (C), whisky (D) and vodka (E).

PROBLEM 15 THE COMPARISON OF FLAME IONIZATION AND ELECTRON CAPTURE DETECTORS IN GLC *(TEST QUESTION)*

INTRODUCTION

Flame ionization and electron capture detectors vary considerably in their specificity and sensitivity, and it is important to appreciate the differences and to be able to select an appropriate detector for a particular task. Refer to 'Gas Liquid Chromatography' (p. 26) for background information and to *Analytical Biochemistry* (§ 3.2.3).

ANALYTICAL PROBLEM

Simple organic acids such as benzoic acid and its derivatives can be separated by gas chromatography as their TMS (trimethylsilyl) esters. A gas chromatographic method for their determination using an OV101 column has been developed. Both FID (flame ionization) and ECD (electron capture) detectors are available, but it still remains to decide which detection system would be the most appropriate for the sample under investigation.

The following questions need to be answered:

A Will both detectors respond to all compounds?

B Is there any difference in the sensitivity of the two detectors?

C Is the use of a single detection system valid for the analysis?

INVESTIGATION

TMS derivatives of the three organic acids listed below, each at two concentrations (10 mg ml^{-1} and 0.01 mg ml^{-1}) and an unknown sample of typical composition were prepared and separated by GLC using two different detection systems, FID and ECD.

Chromatogram	*Analyte*	*Concentration* (mg ml^{-1})	*Detector*
A	Benzoic acid	10.00	FID
B	Benzoic acid	0.01	ECD
C	Chlorobenzoic acid	10.00	FID
D	Chlorobenzoic acid	0.01	ECD
E	Methylbenzoic acid	10.00	FID
F	Methylbenzoic acid	0.01	ECD
G	Sample		FID
H	Sample		ECD

DATA

Figure 10 (a-h) shows the chromatograms for the separations.

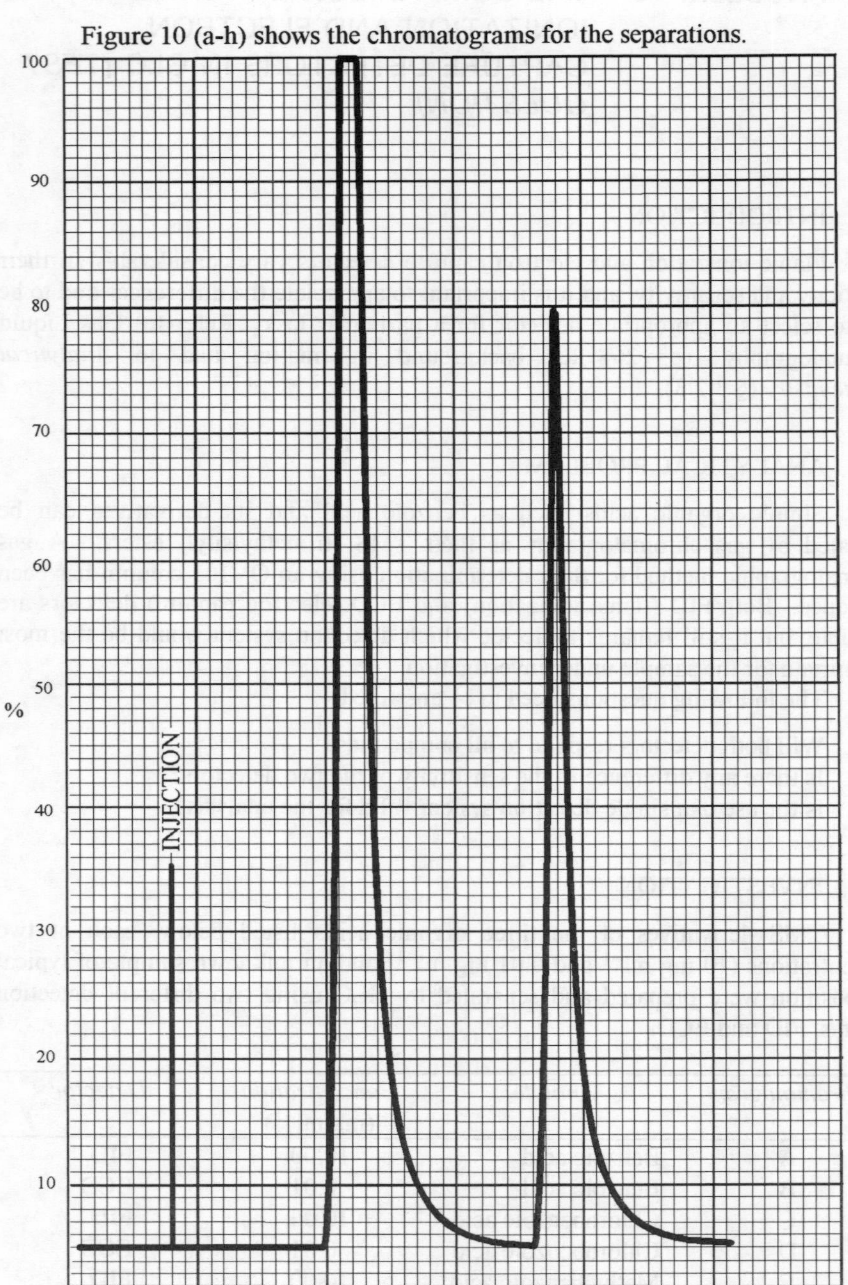

Fig. 10 Chromatograms of (a) Benzoic acid (10 mg ml^{-1}) by flame ionization detection

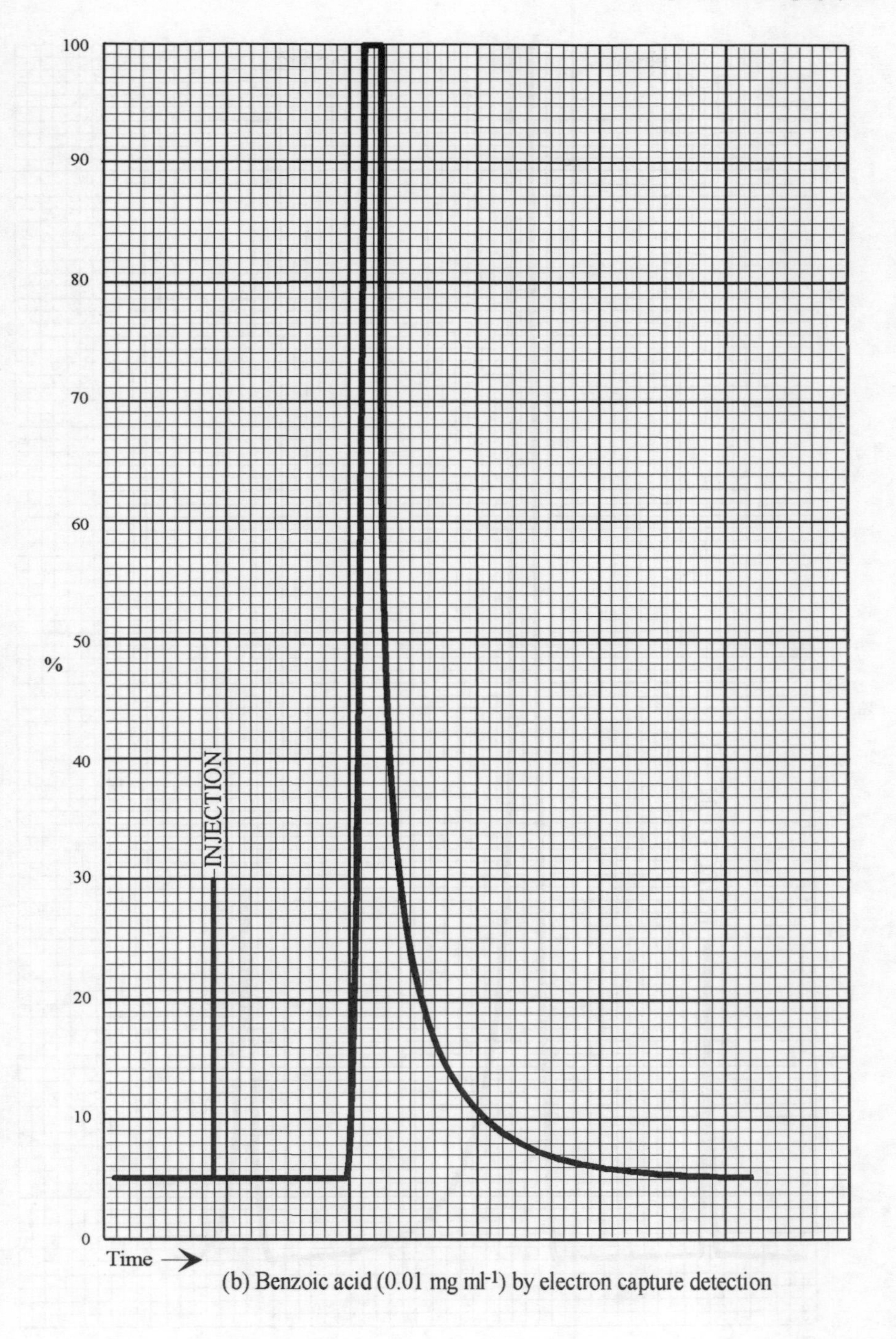

(b) Benzoic acid (0.01 mg ml^{-1}) by electron capture detection

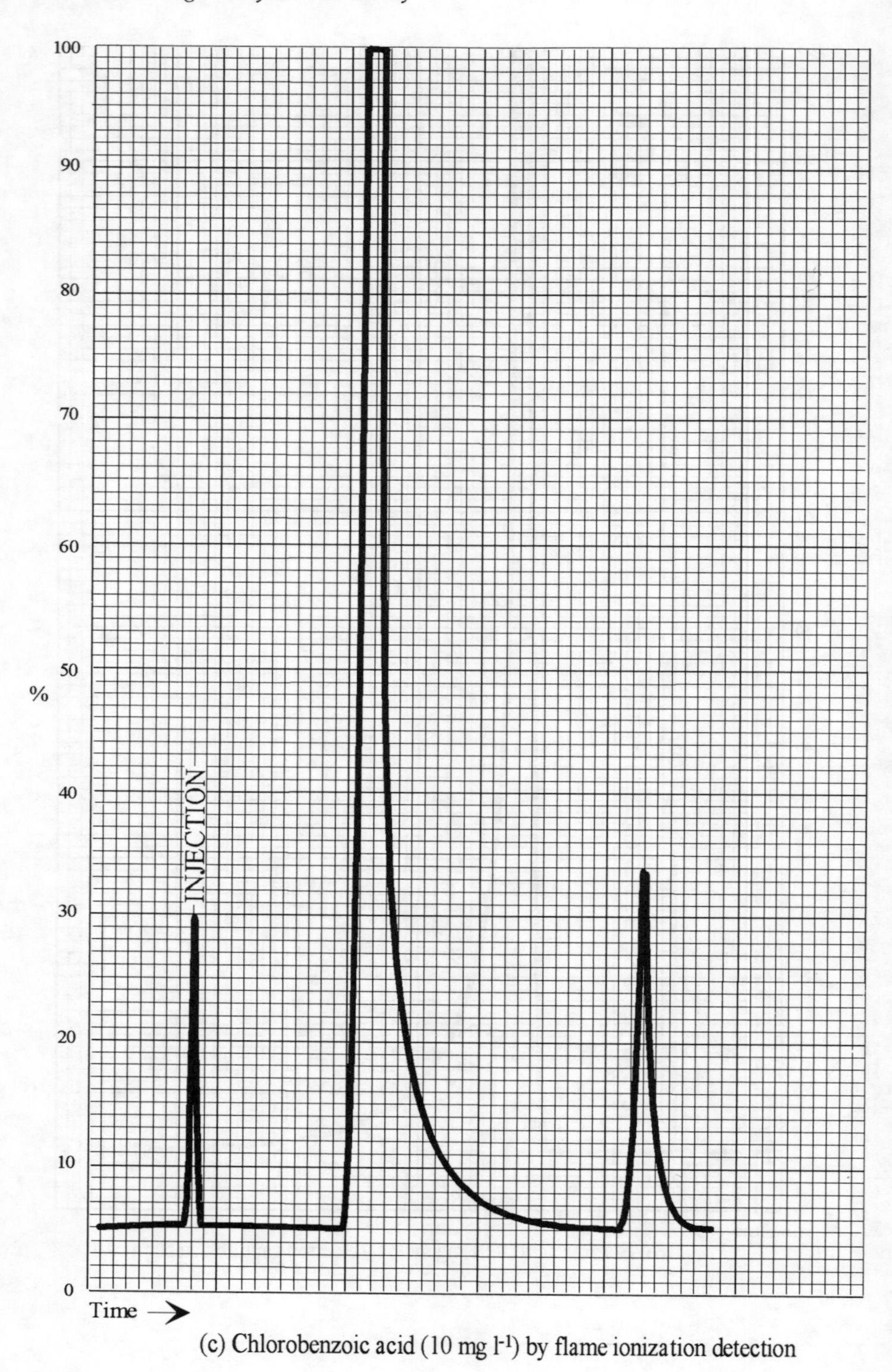

(c) Chlorobenzoic acid (10 mg l^{-1}) by flame ionization detection

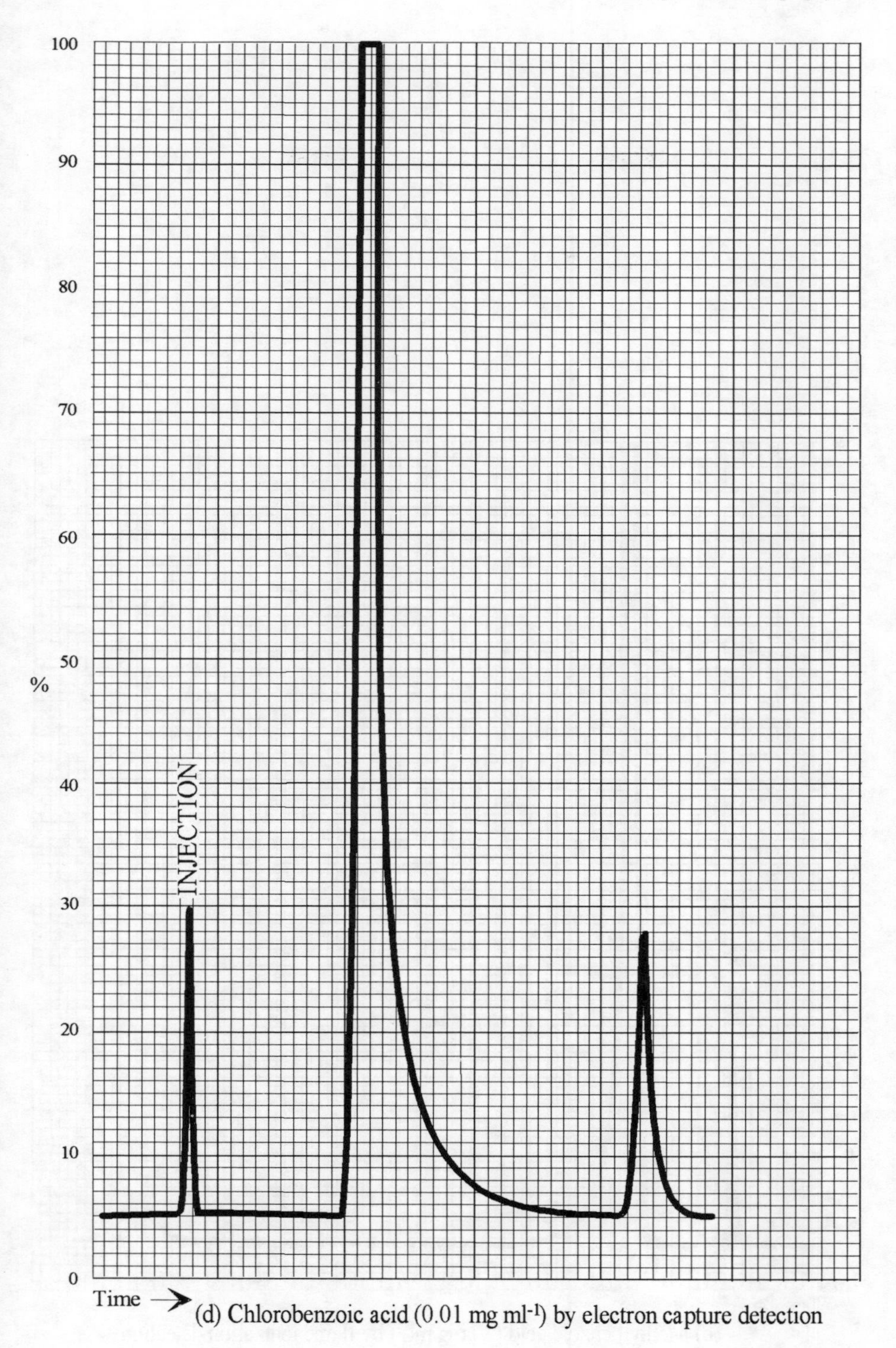

(d) Chlorobenzoic acid (0.01 mg ml^{-1}) by electron capture detection

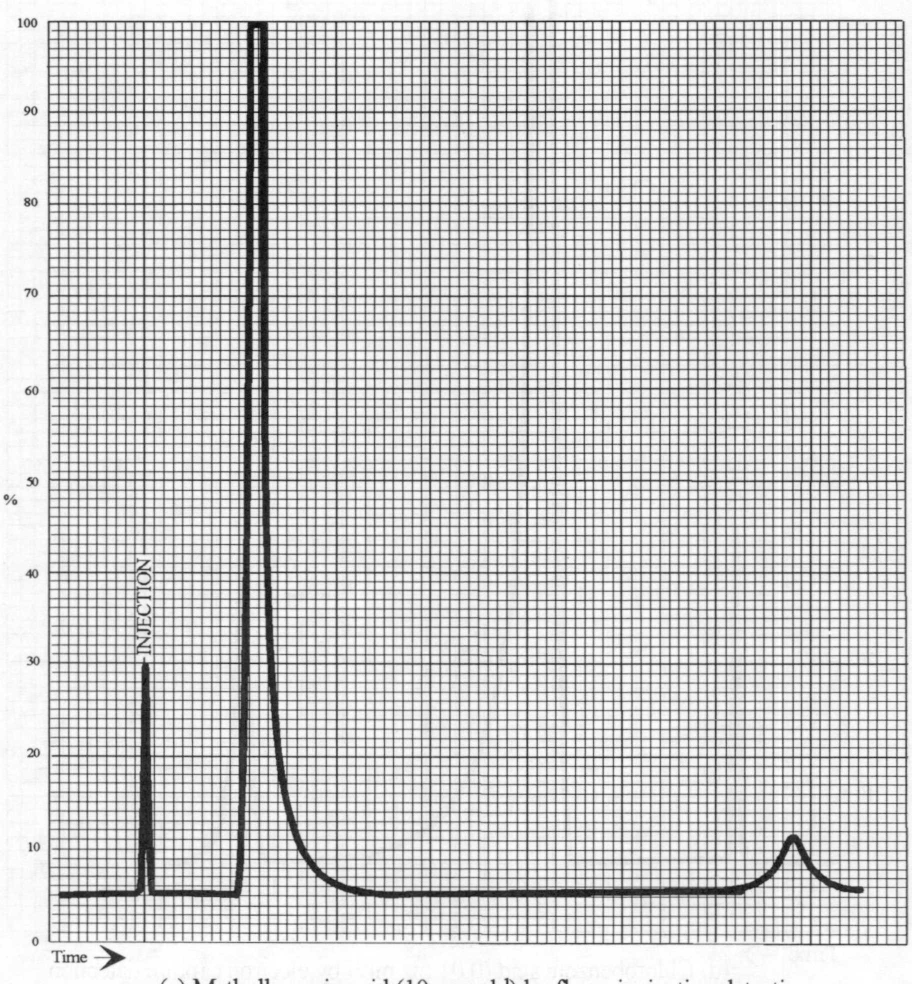

(e) Methylbenzoic acid (10 mg ml^{-1}) by flame ionization detection

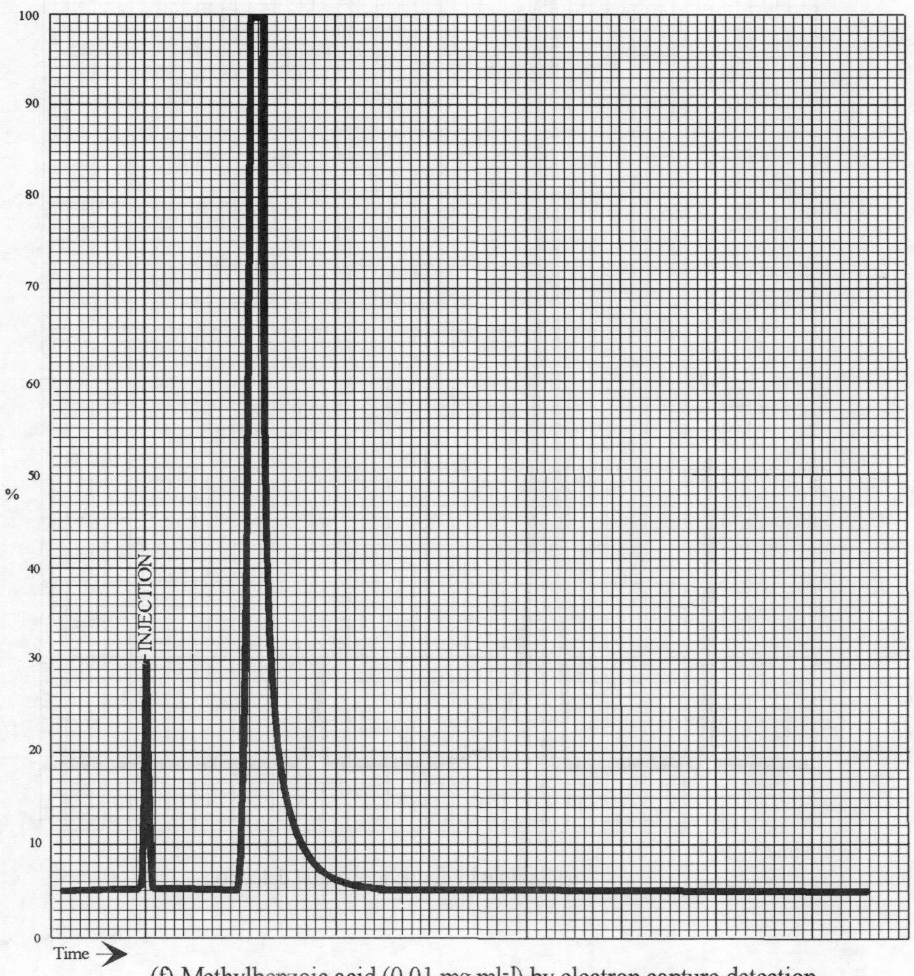

(f) Methylbenzoic acid (0.01 mg ml^{-1}) by electron capture detection

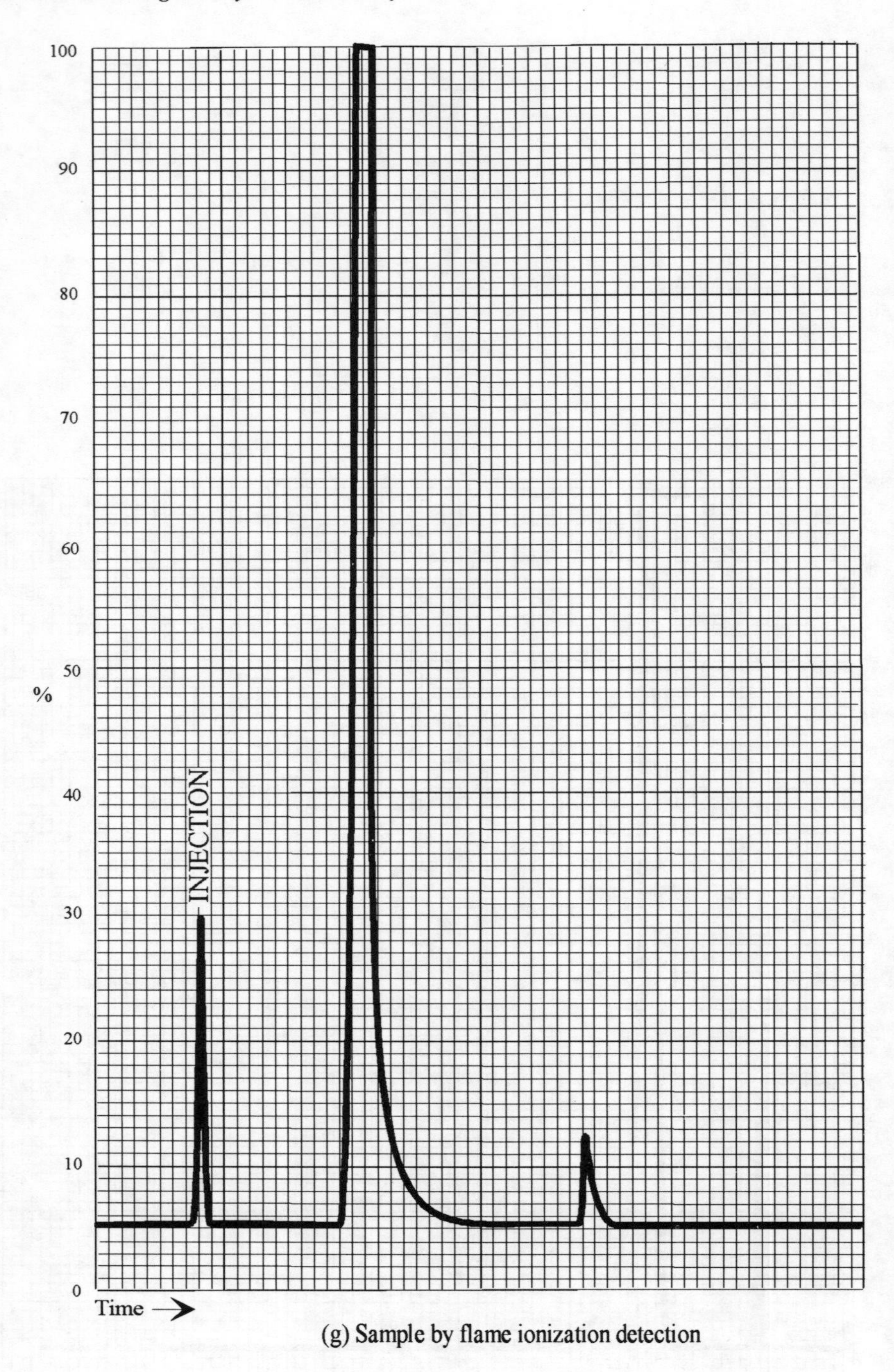

(g) Sample by flame ionization detection

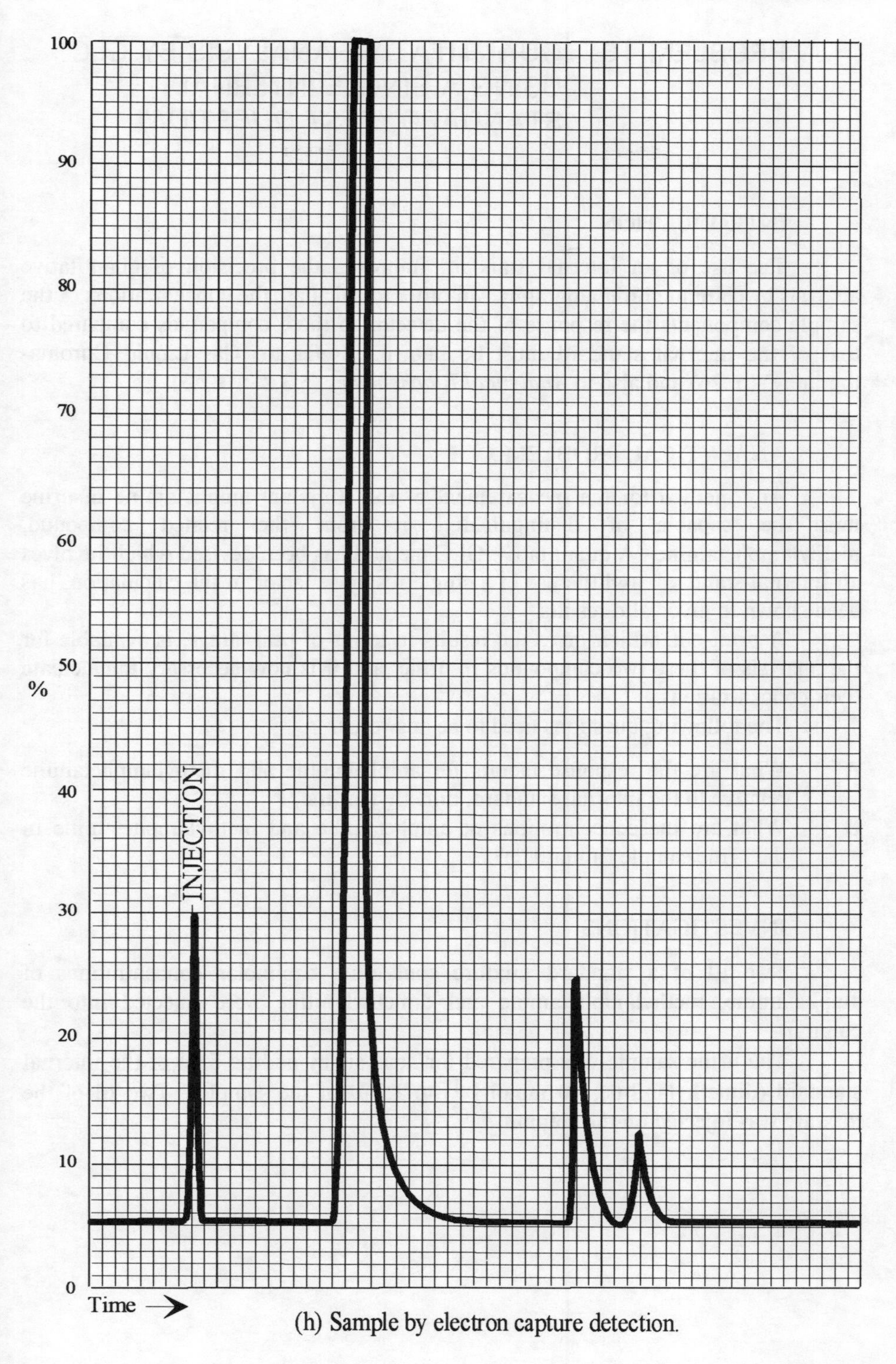

(h) Sample by electron capture detection.

PROBLEM 16 QUANTITATIVE ANALYSIS BY GLC USING A SINGLE INTERNAL STANDARD *(TEST QUESTION)*

INTRODUCTION

The use of an internal standard improves the precision of quantitative analysis by column chromatography. In order to calculate the concentrations of the sample components the response of the detector to these compounds compared to that of the internal standard must be known. Refer to 'Gas Liquid Chromatography' (p. 26) and also to *Analytical Biochemistry* (§ 3.2.2, 3.2.3).

ANALYTICAL PROBLEM

Any method for the measurement of the stimulant amphetamine in urine must be capable of distinguishing it from the related compound, methylamphetamine. A quantitative GLC method has been devised which involves an internal standard, and the use of a single response factor in the calculation, has been shown to give valid results.

A print-out, which gives the results in terms of peak areas, is available for the analysis of these two compounds in urine, and it is now necessary to calculate their concentrations.

The following questions need to be answered:

A What are the response factors for amphetamine and methylamphetamine relative to the internal standard, dimethylaniline?

B What are the concentrations of amphetamine and methylamphetamine in the urine sample (mmol l^{-1})?

INVESTIGATIONS

Two μl of a standard mixture containing equimolar concentrations of amphetamine, methylamphetamine and dimethylaniline were injected into the column.

The urine sample was prepared for analysis by adding 5 μl of the internal standard (dimethylaniline, 10 mmol l^{-1}) to 15 μl of the sample. Two μl of the mixture was injected on the column.

DATA

Analysis of equimolar solutions, and urine compounds

Compound	*Peak area*
Equimolar solutions	
Amphetamine	12
Methylamphetamine	20
Dimethylaniline	16
Urine compounds	
Amphetamine	8
Methylamphetamine	12
Dimethylaniline	16

5 HIGH PERFORMANCE LIQUID CHROMATOGRAPHY

High performance liquid chromatography (HPLC) refers to a standard type of instrumentation for column chromatography which may be used in a variety of modes depending upon the type of stationary phase used. It consists of a sophisticated pump that can deliver the mobile liquid phase at preselected flow rates and at the relatively high pressures necessary for the tightly packed separation columns that are frequently used. A range of different detection systems is available, the most frequently used being absorption spectroscopy but fluorescence and electrochemical detectors are also popular.

HPLC is most frequently used in the reverse phase liquid chromatography mode in which a non-polar stationary liquid phase, immobilized on a particulate supporting medium, is used in conjunction with a polar mobile phase. However, HPLC instrumentation is also used for ion-exchange and gel permeation chromatography.

REVERSE PHASE LIQUID CHROMATOGRAPHY

The versatility of reverse phase HPLC lies mainly in the fact that there is a large number of mobile phase mixtures with varying polarities which can be used. The selection of a suitable mobile phase demands an appreciation of the polarity of the analyte molecules, because a mobile phase of similar polarity will be required. The selected mobile phase must be assessed for its effectiveness and parameters such as resolution index, number of theoretical plates and HETP, should all be determined (see p.26).

GEL PERMEATION CHROMATOGRAPHY

Gel permeation chromatography (GPC) is also known as gel filtration or molecular exclusion chromatography. The gel structure contains pores of varying diameter up to a maximum size for a particular gel. The test molecules are washed through a column of the gel and molecules larger than the largest pores (the exclusion limit) are excluded from the gel structure, and pass quickly through the column in a volume of mobile phase equal to the void volume (V_o). Smaller molecules penetrate the gel to a varying extent depending upon their size, and this retards their progress through the column. Therefore, the compounds are eluted in order of decreasing size.

Thus the greater the RMM of the test substance, the smaller will be the volume required to elute it from the column (V_e) and therefore, the shorter will be the retention time. In order to minimize the effect of fluctuations in flow rate, the ratio of V_e/V_o (often expressed in terms of retention distance) is usually calculated for each substance.

GPC can also be used to determine the relative molecular mass of large molecules such as proteins. If the analysis of both standards and the test is performed at the same time, and under identical conditions, then a plot of the logarithm of RMM against V_e/V_o can be used to determine the unknown RMM.

QUANTITATION

The design of the injection system in HPLC considerably improves the precision of sampling compared with needle syringe injection which is mainly used in GLC. However, all the analytical considerations of external and internal standards are still relevant to HPLC (see p. 54).

PROBLEM 17 THE SELECTION OF A MOBILE PHASE FOR REVERSE PHASE HPLC

INTRODUCTION

Reverse phase chromatography using non-polar ODS (octadecylsilyl) columns is a versatile method for HPLC and in such separations an appropriate mobile phase must be used. In selecting a mobile phase, consideration of the relative polarity of the analytes and potential mobile phases is essential. Refer to 'High Performance Liquid Chromatography' (p. 54) for background information and to *Analytical Biochemistry* (§ 3.2.2).

ANALYTICAL PROBLEM

It is intended to attempt to separate two vitamins, niacin and riboflavin by reverse phase HPLC using an ODS column and a mobile phase of an aqueous solution of either methanol or acetonitrile.

The following question needs to be answered:

- What would be the optimal composition of a mobile phase for the effective separation of these two vitamins?

INVESTIGATIONS

A mixture of the two vitamins was analysed using various aqueous dilutions of both methanol and acetonitrile.

DATA

The chromatograms for the separations are shown in Fig. 11 for:

(a) 25% methanol,
(b) 50% methanol,
(c) 75% methanol,
(d) 10% acetonitrile,
(e) 20% acetonitrile.

SOLVING THE PROBLEM

Only refer to this section when you have made an attempt to answer the question.

1. Measure and tabulate all the retention times.
2. Calculate the resolution index for each separation.

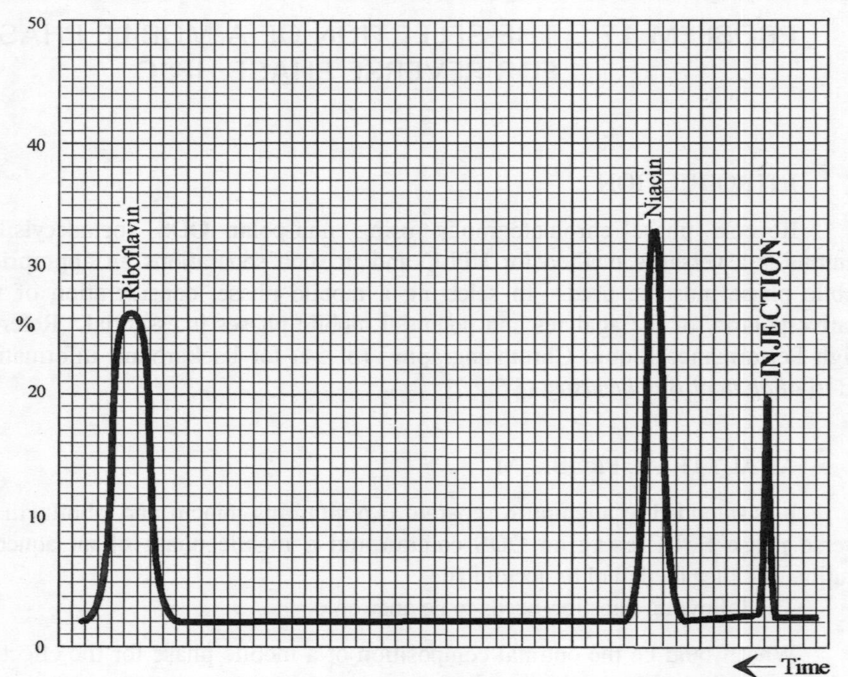

Fig. 11 Selection of mobile phase in HPLC for (a) 25% methanol in water

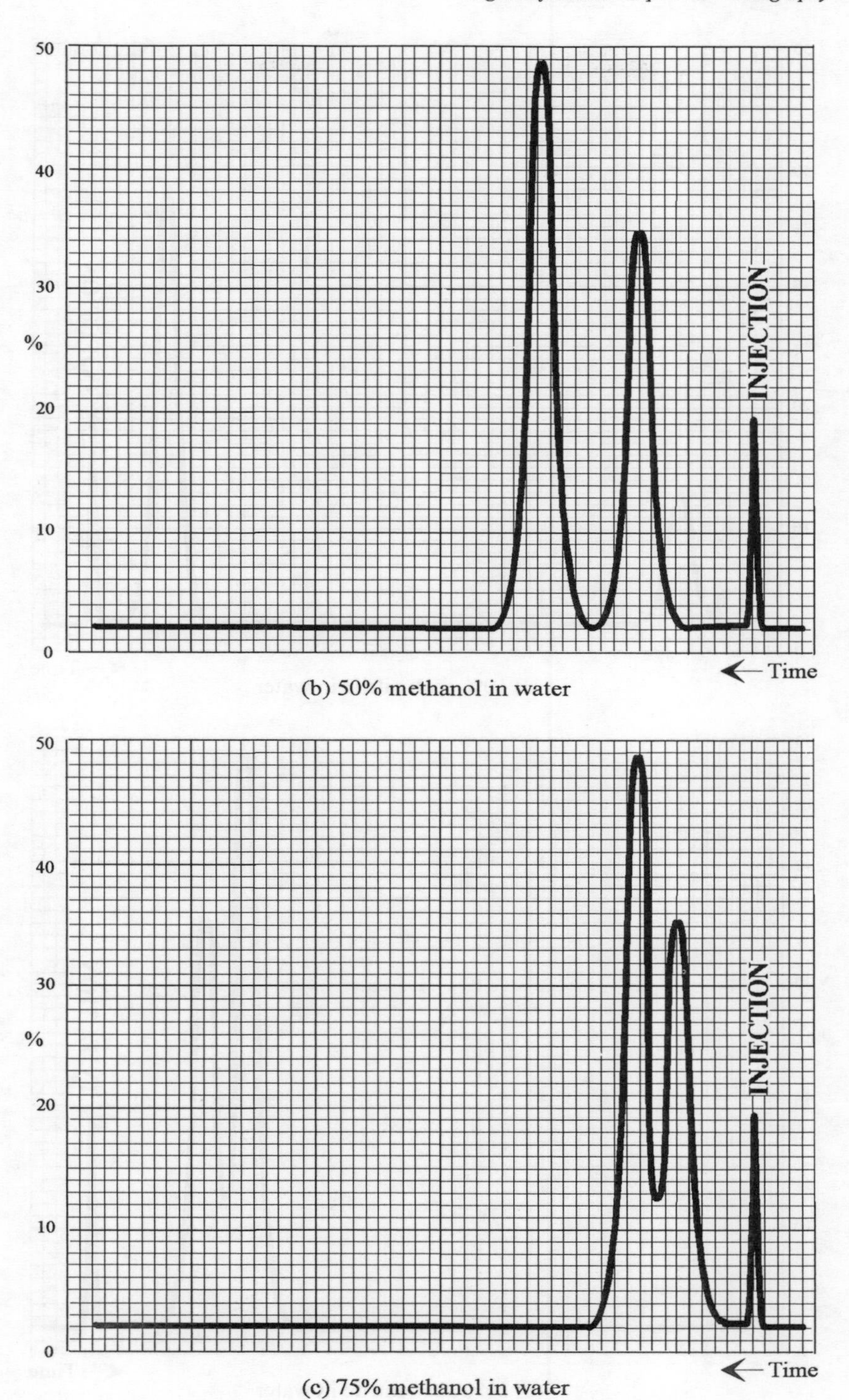

(b) 50% methanol in water

(c) 75% methanol in water

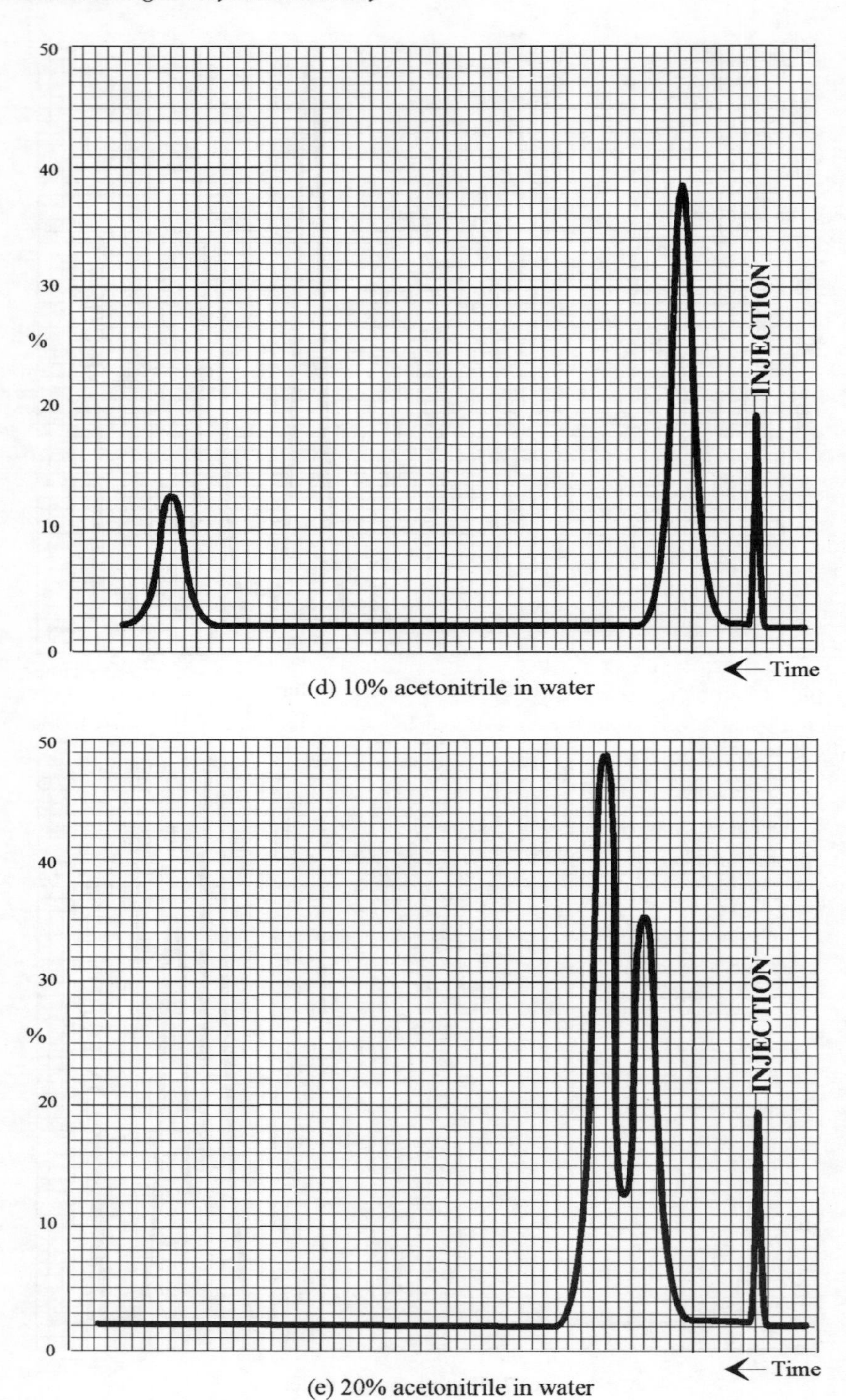

(d) 10% acetonitrile in water

(e) 20% acetonitrile in water

PROBLEM 18 QUANTITATIVE ANALYSIS BY ION-EXCHANGE CHROMATOGRAPHY USING A SINGLE INTERNAL STANDARD

INTRODUCTION

The use of an internal standard improves the precision of quantitative analysis by column chromatography. In order to calculate the concentration of the sample components the response of the detector to these compounds compared to that of the internal standard must be known. These values are normally called response factors but when colorimetric detection is employed, as in the amino acid analyser with ninhydrin reagent, the term colour value may be used. Refer to 'Gas Liquid Chromatography' for background information on response factors (p. 26) and also to *Analytical Biochemistry* (§ 3.2.2, 10.6.1 and 10.6.3).

ANALYTICAL PROBLEM

It is necessary to determine the concentration of three amino acids in a sample of hydrolysed barley from a print-out, giving results in terms of peak areas, which was obtained from an amino acid analyser. An internal standard of nor-leucine was used in the analysis.

The following questions need to be answered:

A What is the response factor for each amino acid relative to the internal standard?

B What is the concentration of each amino acid in the sample (mmol l^{-1})?

INVESTIGATIONS

Ten μl of a standard mixture of amino acids, including the internal standard nor-leucine, each at a concentration of 2.0 mmol l^{-1} were applied to the column and the appropriate separation programme was run.

Twenty μl of nor-leucine (2 mmol l^{-1}) was added to 40 μl of the sample and 10 μl of the mixture was applied to the column and separated as before.

DATA

Analysis of the standard mixture and barley sample

Amino acid	*Peak area*	
	Standard mixture	*Barley sample*
Histidine	30	18
Phenylalanine	20	20
Tyrosine	18	10
Nor-leucine	22	16

SOLVING THE PROBLEM

Only refer to this section when you have made an attempt to answer the question.

1. Calculate the response factor for each of the three amino acids relative to nor-leucine using the peak areas from the standard mixture results. Remember that the concentration for each standard is the same.

2. Calculate the concentration of each amino acid in the sample using the previously calculated response factors and knowing the concentration of the internal standard nor-leucine.

PROBLEM 19 THE DETERMINATION OF RELATIVE MOLECULAR MASS USING GEL PERMEATION CHROMATOGRAPHY

INTRODUCTION

Gel permeation chromatography as well as proving a valuable preparative and quantitative technique, will also permit the determination of relative molecular mass (RMM) under certain conditions. The correct gel must be used and the column calibrated with proteins of known RMM. Refer to 'High Performance Liquid Chromatography' (p. 54) and also to *Analytical Biochemistry* (§ 3.4.2, 11.3.6).

ANALYTICAL PROBLEM

An enzyme has been isolated from plant tissue and its relative molecular mass must be determined. It can be separated from other proteins by gel permeation chromatography, being detected in the effluent by monitoring the absorbance at 280 nm. A range of proteins of known RMM are available to calibrate the selected column.

The following question must be answered:

- What is the RMM of the sample enzyme?

INVESTIGATIONS

The five reference compounds and the sample enzyme were separated using a gel permeation medium with an exclusion limit of 1.5×10^6 da.

DATA

The chromatograms for the separations are shown in Figure 12 (a-f).

Chromatogram	*Reference protein or dye*	*RMM* (da)
A	Alcohol dehydrogenase	1.5×10^5
B	Bovine albumin	6.6×10^4
C	α-amylase	2.0×10^5
D	Cytochrome *c*	1.24×10^4
E	Blue dextran	2.0×10^6
F	Unknown enzyme	?

SOLVING THE PROBLEM

Only refer to this section when you have made an attempt to answer the question

1. For each reference protein measure the retention distance (V_e) and calculate the ratio V_e/V_o (relative elution volume), the void volume being given by the retention distance of blue dextran.

2. Plot the relative elution volume for each reference against $\log_{10}$ RMM and use the graph to determine the RMM of the sample enzyme.

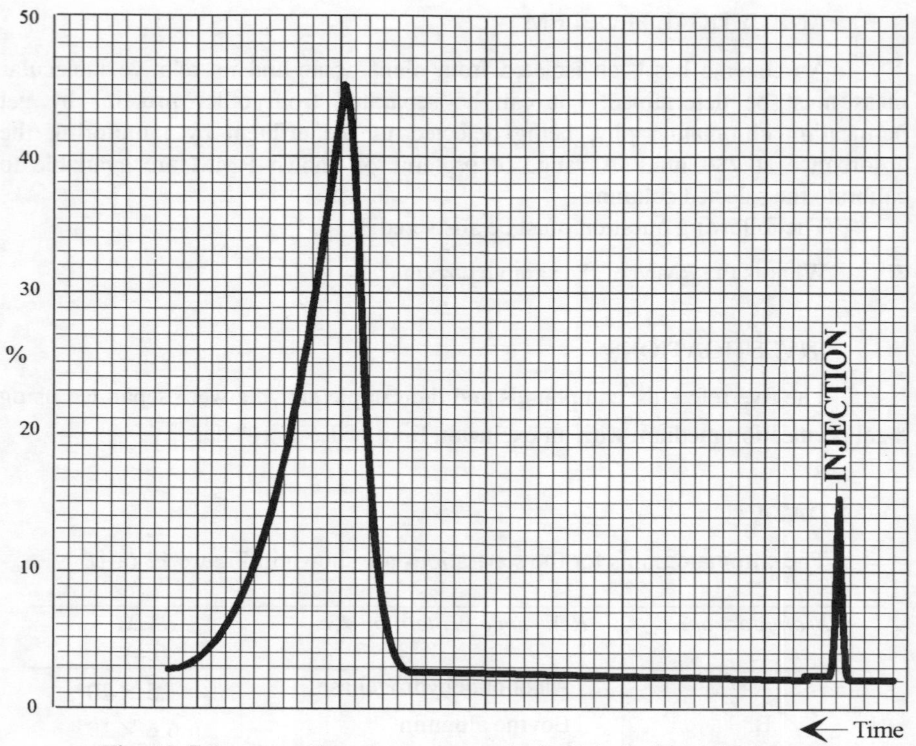

Fig. 12 Gel permeation chromatography for (a) alcohol dehydrogenase

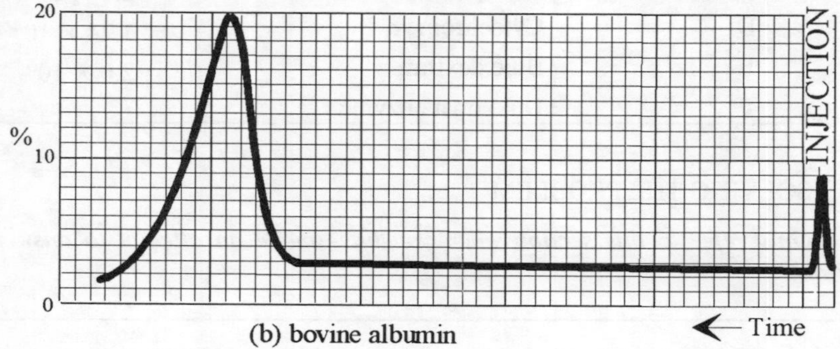

(b) bovine albumin

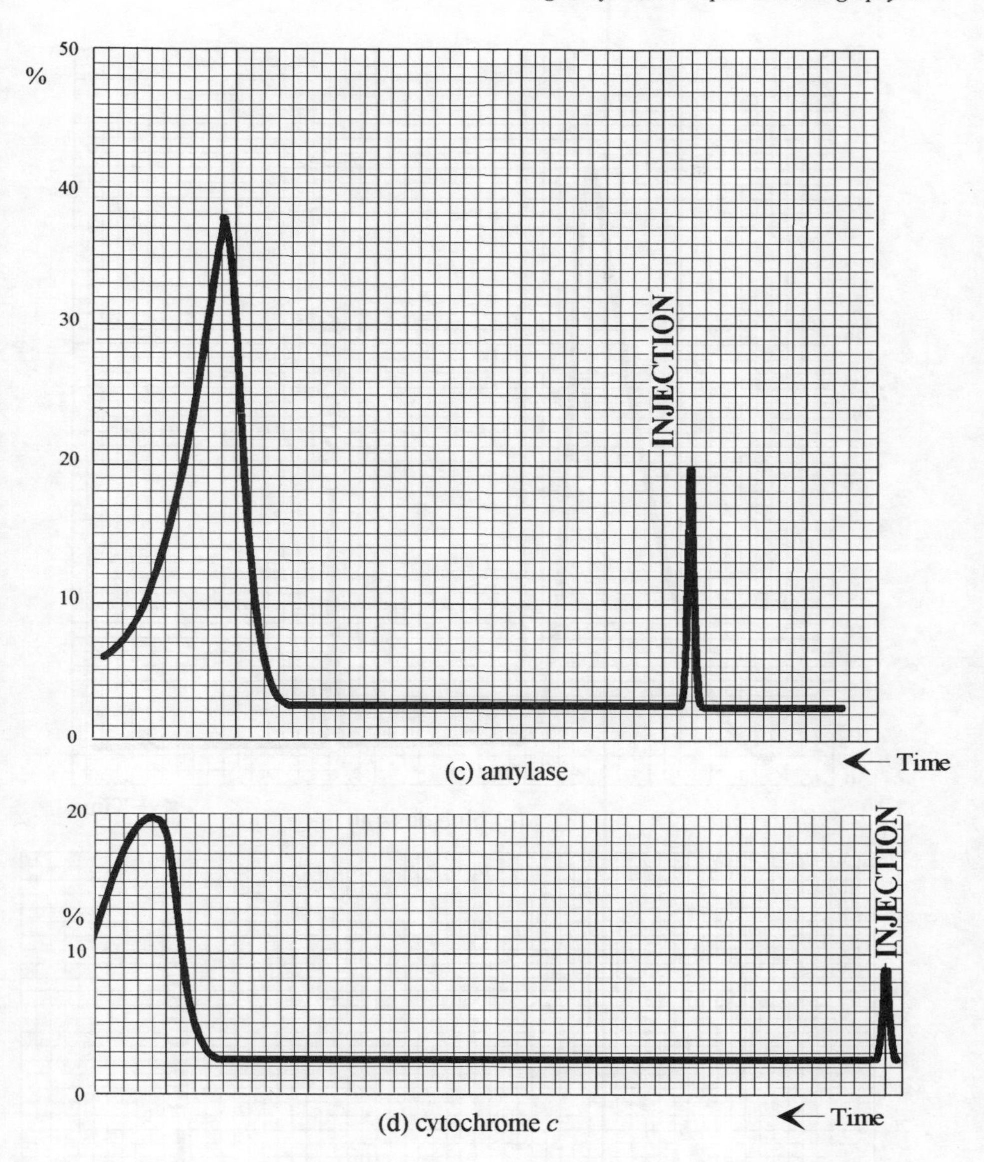

(c) amylase

(d) cytochrome *c*

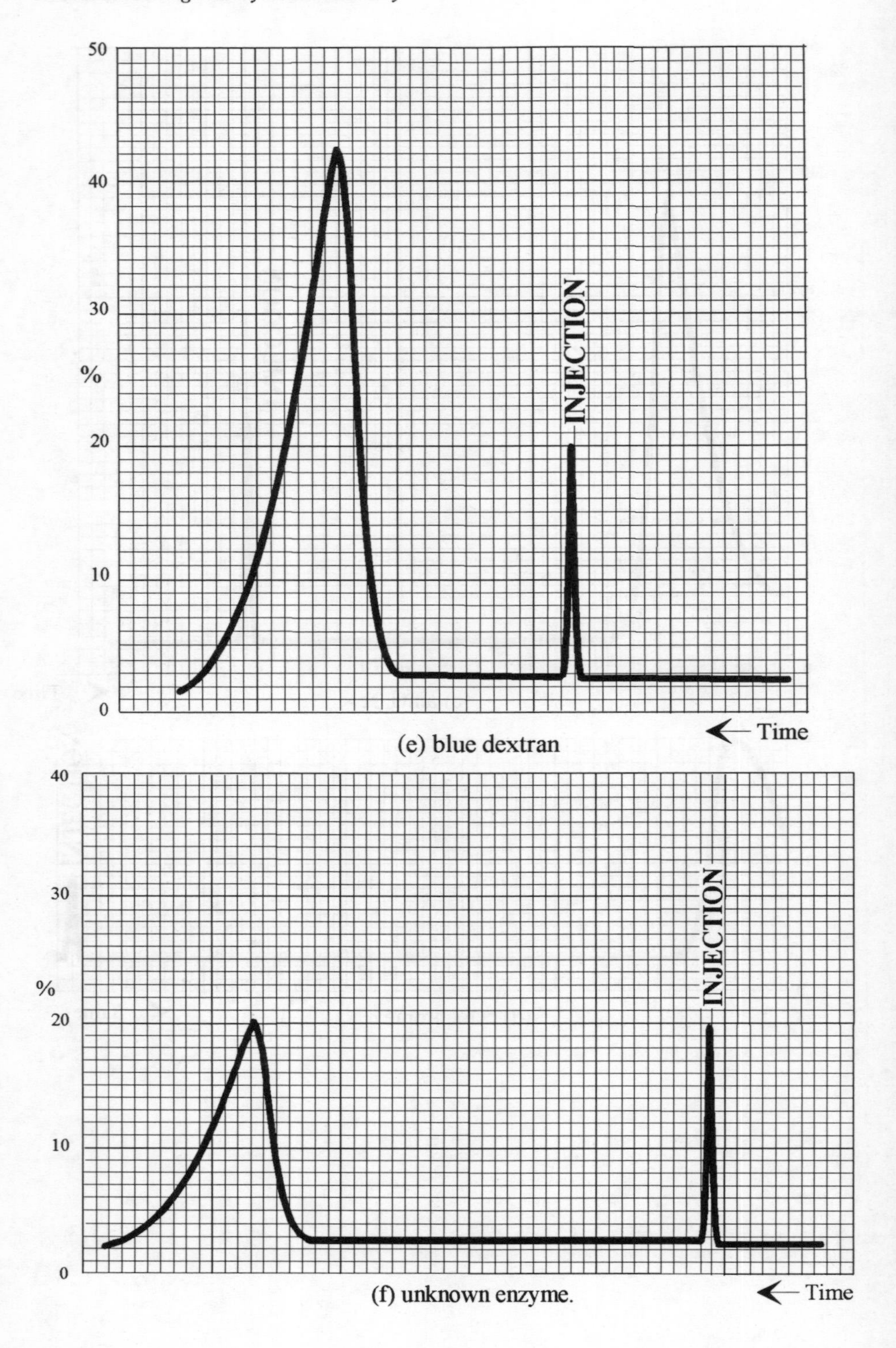

(e) blue dextran

(f) unknown enzyme.

PROBLEM 20 SELECTION OF THE OPERATING VOLTAGE FOR ELECTROCHEMICAL DETECTION IN HPLC

INTRODUCTION

In developing any HPLC method, the performance of the detector must be assessed and the optimum operating conditions determined. Electrochemical detectors are gaining popularity and are applicable for the detection of any compound which undergoes an electrochemical reaction at an electrode. For each compound there is a minimum voltage below which the reaction will not occur; above this value a current will flow between the electrodes, and it will increase up to a maximum as the voltage is increased. The voltage which produces the maximum current is characteristic of the compound, but the actual current which flows is proportional to the amount of analyte present. Refer to 'High Performance Liquid Chromatography' for background information (p. 54) and also to *Analytical Biochemistry* (§ 3.2.2, 4.4.2).

ANALYTICAL PROBLEM

A method is to be developed for the separation of three derivatives of benzoic acid by reverse phase HPLC using electrochemical detection. The optimum operating conditions need to be determined.

The following question needs to be answered:

- What is the optimum operating voltage for the detector to be able to detect all three compounds simultaneously?

INVESTIGATIONS

Five ng amounts of each compound in solution were separately injected onto the column with the detector set at a preselected voltage and the resulting peak heights were recorded from the chromatogram. The process was repeated with the detector set at a range of voltages between 800 and 1200 mV.

DATA

Voltage (mV)	*Compound peak height* (mm)		
	p-amino benzoic acid	*m*-hydroxy benzoic acid	*o*-amino benzoic acid
800	0	0	2
850	1	0	2
900	2	1	3
950	15	10	5
1000	25	20	10
1050	32	34	15
1100	40	42	36
1150	39	51	36
1200	38	52	38

SOLVING THE PROBLEM

Only refer to this section when you have made an attempt to answer the question.

1. Plot the data for all the compounds on the same graph and determine a voltage which gives the maximum sensitivity for all three compounds.

PROBLEM 21 QUANTITATIVE ANALYSIS BY HPLC USING AN EXTERNAL STANDARD *(TEST QUESTION)*

INTRODUCTION

The accuracy and precision associated with the use of an injection loop in HPLC makes the use of an external standard much more reliable than some other methods of injecting the sample. However, it does not remove the possibility of analytical conditions varying between subsequent injections and the consequent implications on the quantitative aspects of the analysis. Refer to 'High Performance Liquid Chromatography' (p. 54) and also *Analytical Biochemistry* (§ 3.2.2).

ANALYTICAL PROBLEM

The validity of a reverse phase HPLC method for the quantitation of caffeine using an external standard has been demonstrated. This method has been used to analyse samples of normal and decaffeinated coffee and the concentrations now need to be calculated.

The following question needs to be answered:

- What is the concentration of caffeine in each sample (mmol l^{-1})?

INVESTIGATIONS

Three dilutions of a standard caffeine solution were prepared and analysed successively using an ODS column, acetonitrile as the mobile phase and detection at 254 nm.

Samples of normal and decaffeinated instant coffee were injected and the chromatograms recorded.

DATA

The chromatograms for the separations are shown in Figure 13 (a-e).

Chromatogram	*Sample*
A	Caffeine – 0.5 mmol l^{-1}
B	Caffeine – 1.0 mmol l^{-1}
C	Caffeine – 2.0 mmol l^{-1}
D	Coffee sample – normal
E	Coffee sample – decaffeinated

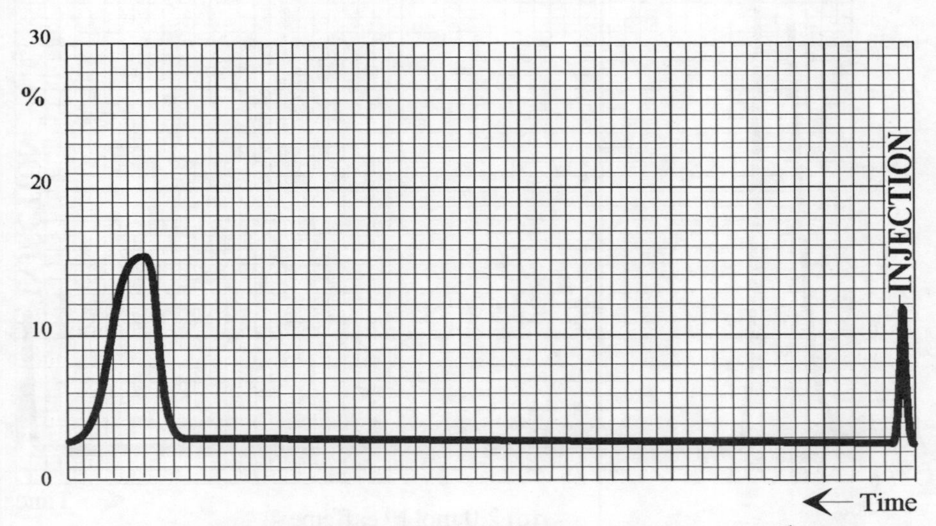

Fig. 13 Caffeine determination by HPLC for (a) 0.5 mol l caffeine

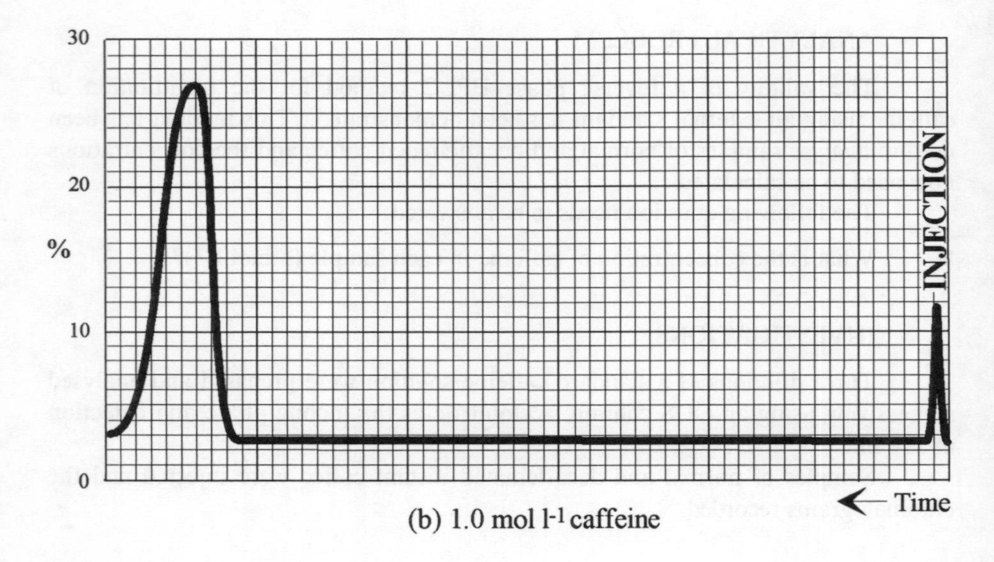

(b) 1.0 mol l^{-1} caffeine

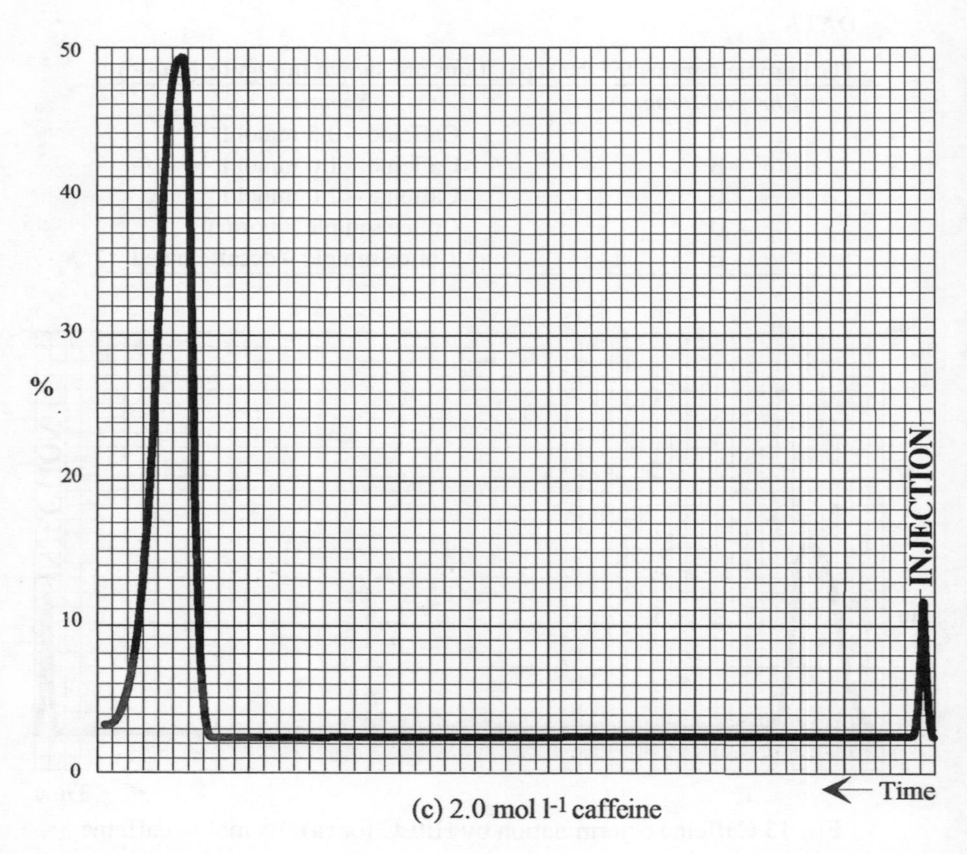

(c) 2.0 mol l^{-1} caffeine

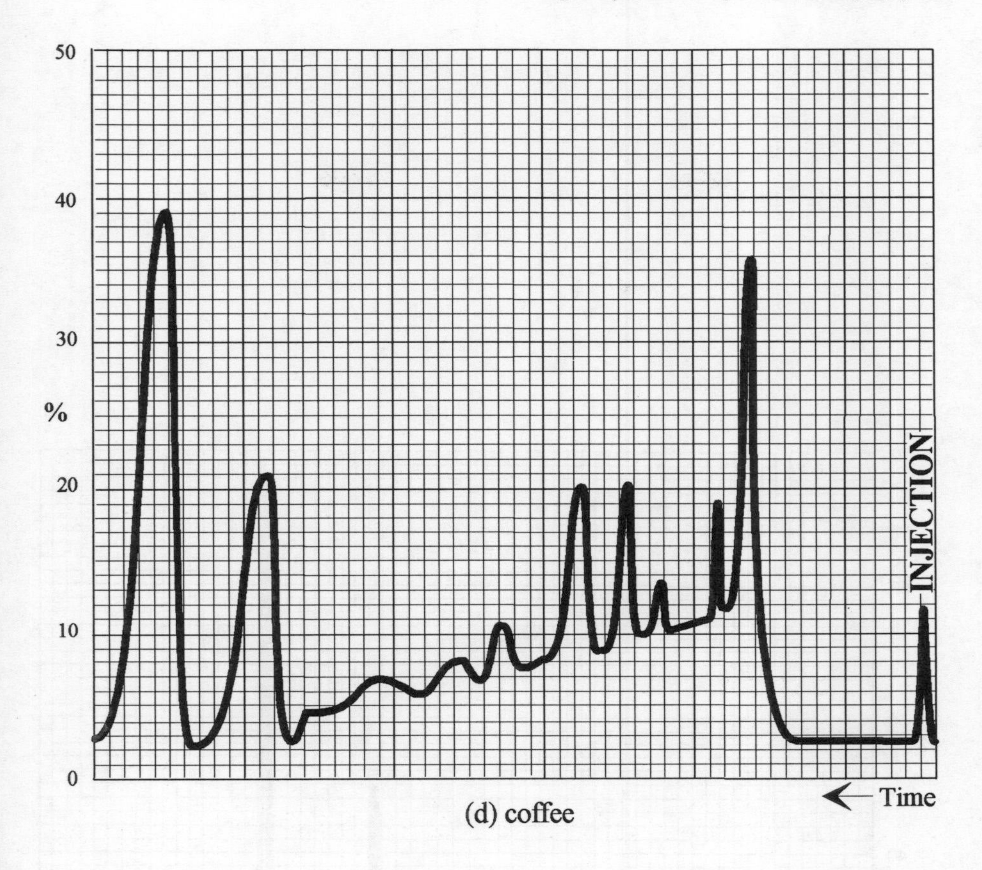

(d) coffee

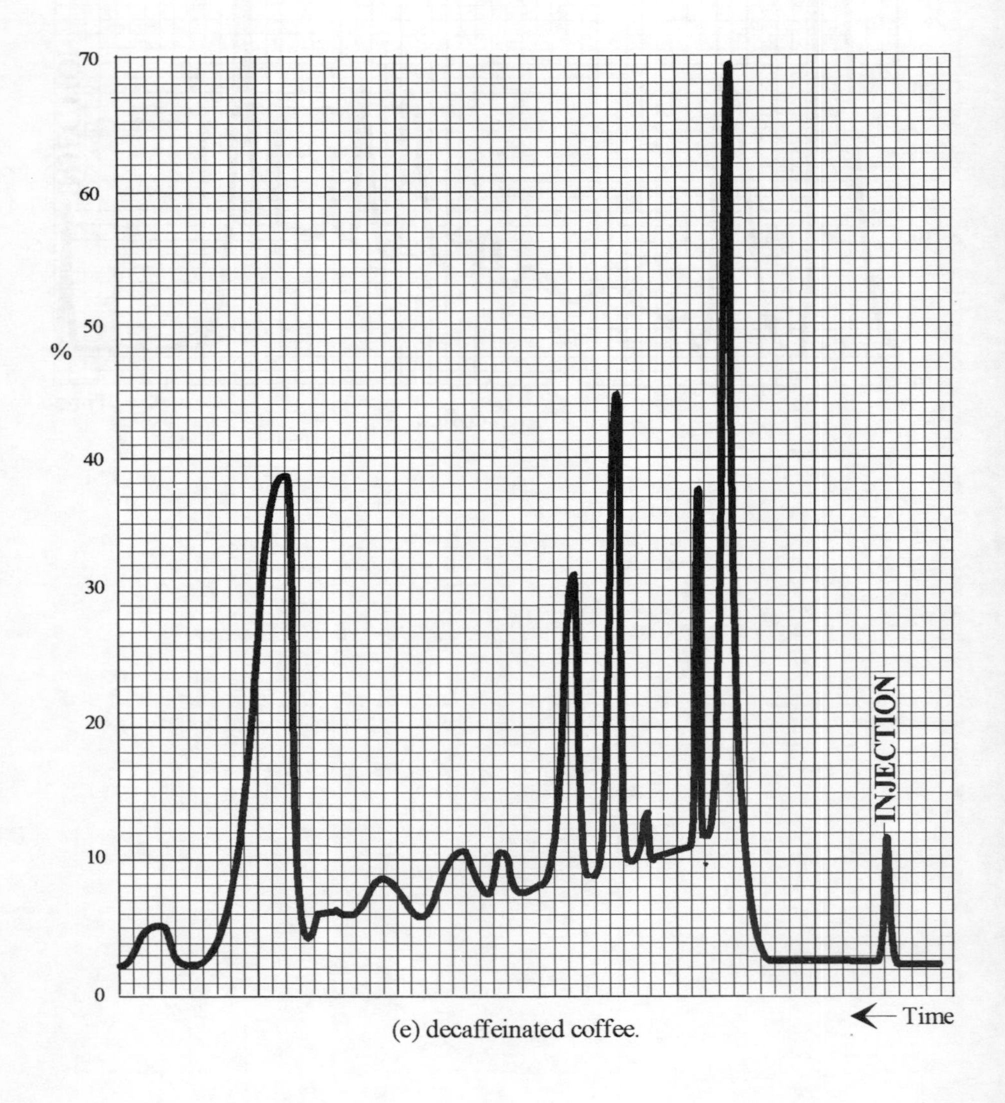

(e) decaffeinated coffee.

PROBLEM 22 ASSESSMENT OF THE EFFICIENCY OF SEPARATION BY HPLC *(TEST QUESTION)*

INTRODUCTION

It is essential to be able to assess the efficiency of any column chromatographic process and this usually involves the determination of resolution index and the number of theoretical plates. This may be used for the purpose of optimizing chromatographic conditions, or for monitoring column performance over a period of time. Refer to 'Gas Liquid Chromatography' (p. 26) and also to *Analytical Biochemistry* (§ 3.2.2).

ANALYTICAL PROBLEM

An HPLC method which involves internal standardization is routinely used for the analysis of cyclobarbitone. In order to detect any slight deterioration in column performance with continuous use, the efficiency of the separation process is regularly monitored. The chromatogram which was obtained for this purpose is available for determination of the relevant parameters.

The following questions need to be answered:

A Is the resolution index for the two compounds acceptable?

B What is the efficiency of the separation for cyclobarbitone expressed as the number of theoretical plates?

INVESTIGATIONS

A mixture of the internal standard, diethylbarbituric acid, and cyclobarbitone was analysed using the standard method.

DATA

The chromatogram for the separation is shown in Figure 14.

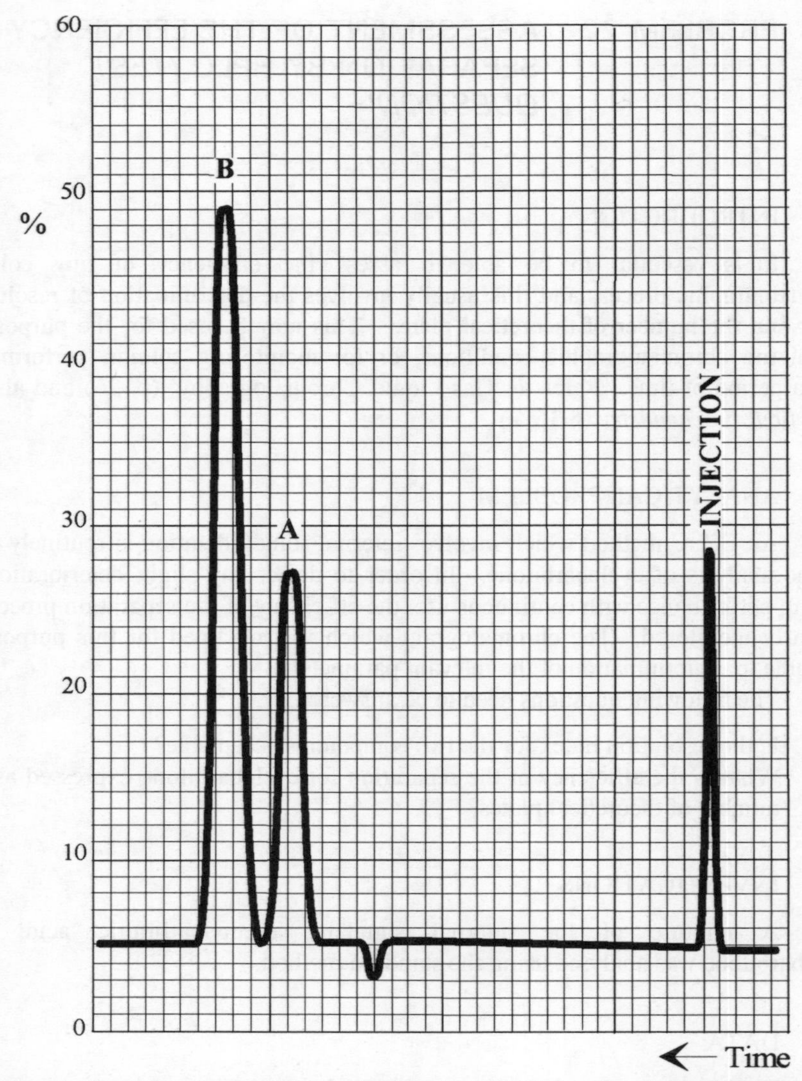

Fig 14 Chromatogram of diethylbarbituric acid (A) and cyclobarbitone (B).

6
IONIC SEPARATIONS

ION-EXCHANGE CHROMATOGRAPHY

An ion-exchange medium consists of an insoluble, porous matrix containing large numbers of a particular ionic group which are capable of binding ions of an opposite charge from the surrounding solution. Hence, a cation exchange medium contains fixed anions, and the mobile ions can be any cations from the solution. Ion-exchange column chromatography is a displacement technique in which ions in either the sample or the eluting buffer displace the existing mobile ions associated with the fixed ions on the medium. For exchange to take place the fixed group must be ionized, a feature which is related to its strength (or ability to dissociate) and is pH dependent.

When developing an ion-exchange separation method careful consideration must be given to the pH effects on the ion-exchange medium and the test molecules. They must both be ionized at the selected pH and carry opposite charges in order for binding to occur. Subsequently the pH of the eluting buffer must be carefully selected to permit sequential solution of the different components of the sample.

ELECTROPHORESIS

Electrophoresis is the term given to the migration of charged particles under the influence of a direct electric current, and separation depends upon the relative mobilities of ions under identical electrical conditions.

Samples are normally applied as a small zone, or band, to a supporting medium which is soaked in a buffered electrolyte and which forms a bridge between the electrodes. The most frequently used media include cellulose acetate membrane and several gels, such as agarose and polyacrylamide. These gels are fragile and demand careful handling. A problem with some supporting media is the phenomenon of electroendosmosis in which the buffer itself moves due to an electrophoretic effect and hence masks the movement of the test substances to some extent.

The major factors in electrophoretic separations are the charge carried by the molecule and the voltage supplied. The nature of the charge will determine the direction of migration, while the magnitude of the charge will determine the relative velocity. The charge carried by a molecule is primarily affected by the pH of the buffered electrolyte which must be carefully chosen. The size of the molecule becomes an additional factor in separations using polyacrylamide and agarose gels in which the movement of the larger molecules is hindered by the pore structure of the gel.

IONIC PROPERTIES OF AMPHOLYTES

Ampholytes, of which amino acids are important examples, contain both acidic and basic groups, and as a result can act both as weak acids and weak bases. Their behaviour is termed amphiprotic because they can either accept or donate a proton. The uncharged form of an amino acid (R.NH_2.COOH) carries one negative and one positive charge and as a result shows no net charge. This is known as the dipolar form or 'zwitterion'.

The dissociated and undissociated forms of each group exist in equilibrium with each other, and the position of the equilibrium may be expressed in terms of the equilibrium (dissociation) constant K, often termed K_a because it refers to the dissociation of groups which liberate protons, i.e. acids. The actual values for K_a are often very small and are conventionally expressed as the negative logarithm of the value, a term known as the pK_a value.

$$pK_a = -\log K_a$$

The concentration of hydrogen ions liberated by the dissociation of an acid is related to the dissociation constant for that acid and this relationship can be expressed by the Henderson-Hasselbalch equation:

$$\text{pH} = pK_a + \log \frac{[\text{salt}]}{[\text{acid}]}$$

where the square brackets indicate the molar concentration of the named substance.

The ionization of an amino acid is most easily demonstrated in a titration curve which can be prepared by titrating a solution of the amino acid in the fully protonated form with a solution of sodium hydroxide and plotting the amount of alkali added against the resulting pH of the solution. The titration curve for a simple amino acid will show two regions where the addition of alkali results only in a small change in the pH value of the solution. The first end-point in such a titration is due to the carboxyl group and the halfway point in that section of the curve, i.e. when the acid group is half neutralised and the acid concentration equals the salt concentration, will give the pK value for that group.

The overall charge carried by an amino acid depends upon the pH of the solution and the pK_a values for ionizable groups present. If the pH is greater than the pK_a value for a group, a proton will be lost and the molecule will carry a negative charge. The iso-ionic point of a molecule is the pH at which the number of negative charges due to proton loss equals the number of positive charges due to proton gain and the zwitterionic form predominates. The iso-electric point (pI) is the pH of the solution at which the molecules show no migration in an electric field and it can be determined experimentally by electrophoresis, and for simple molecules like amino acids it is equal to the iso-ionic pH.

ISO-ELECTRIC FOCUSING

If electrophoresis is performed in a pH gradient rather than at a fixed pH, molecules will migrate to the position in the supporting medium where the pH in the gradient is the same as their pI value, and remain there as long as the gradient and voltage are maintained.

pH gradients are usually produced from a mixture of polyamino–polycarboxylic acids with a range of pI values, and a pH gradient is formed which spans the range of the pI values.

Iso-electric focusing provides a very sensitive method for the separation of mixtures of ampholytes such as proteins, and at the same time gives information on the ionic properties of the compounds in the form of their pI values.

PROBLEM 23 DETERMINATION OF pK_a VALUES BY TITRATION

INTRODUCTION

Monitoring the pH of an amino acid solution while titrating its ionizable groups can give valuable information about the ionic properties and the chemical nature of the compound, as well as quantitative information. The use of an autotitrator coupled with a glass electrode gives a trace which is a plot of the pH of the solution against the volume of titrant added. Refer to 'Ionic Separations' (p. 73) for background information and also to *Analytical Biochemistry* (§ 4.1.2, 10.1.3).

ANALYTICAL PROBLEM

Information is required on the structure and ionic properties of an amino acid using the information gained by titration with sodium hydroxide.

The following questions need to be answered:

A How many ionizable groups does the amino acid possess?

B What are the pK_a values for those groups?

C What can be deduced regarding the chemical nature of the amino acid?

D What is the relative molecular mass of the amino acid?

INVESTIGATIONS

0.05 g of the amino acid was weighed out and dissolved in distilled water. The minimum volume of molar hydrochloric acid was added to bring the pH down to below 2.0.

The resulting solution was titrated in an autotitrator using a sodium hydroxide solution (0.1 nmol l^{-1}) and the titration curve of the amino acid produced over the pH range 1–12.

DATA

The titration curve for the amino acid is shown in Figure 15.

SOLVING THE PROBLEM

Only refer to this section when you have made an attempt to answer the question.

1. Examine the titration curve and deduce the number of ionizable groups from the shape of the curve.
2. From the central section of the trace (the easiest to identify) determine the volume of titrant required to neutralize one group.
3. Determine the pK_a value for this group.

   NB. The pH of the solution when half the volume of titrant required to neutralize a group has been added, will give the pK_a value for that ionizable group.
4. Using the same value for titration volume, determine the pK_a values for the other groups in the amino acid.
5. Calculate the relative molecular mass of the amino acid from the equivalent volume of sodium hydroxide required to neutralize the 0.05 g of amino acid used in the titration.

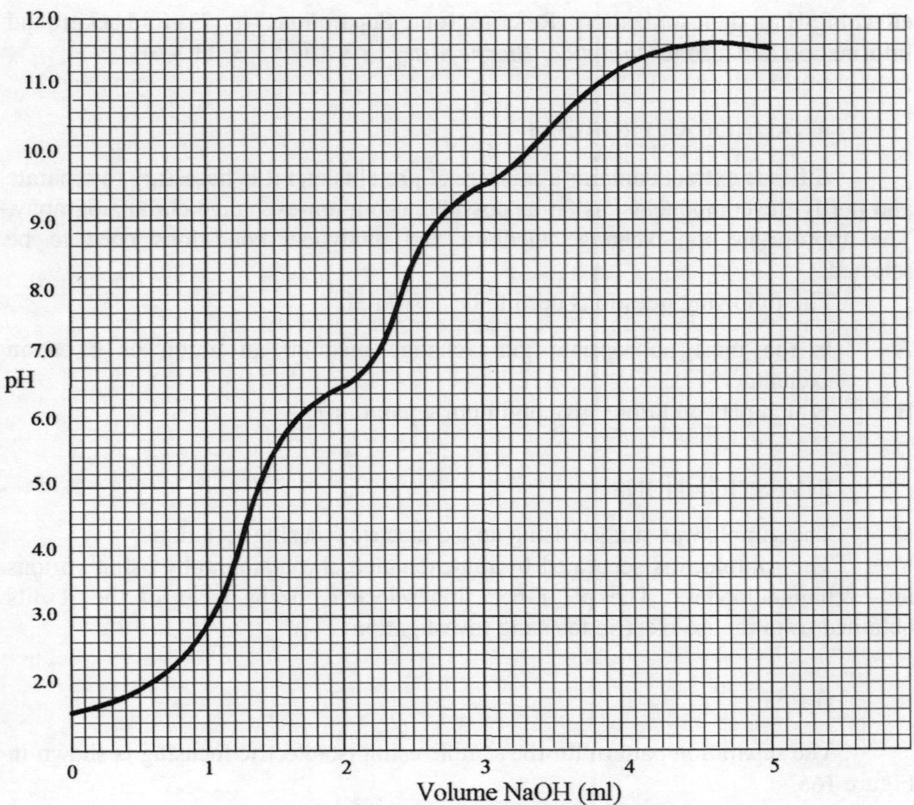

Fig. 15 Titration curve of the test amino acid.

PROBLEM 24 THE SELECTION OF SEPARATION CONDITIONS FOR ION-EXCHANGE CHROMATOGRAPHY

INTRODUCTION

Ion-exchange chromatography demands that the separation medium (gel or resin) must be ionized and that for an effective separation process to occur, the test molecules must also be ionized and carry an opposite charge to the fixed ions in the medium. By using an appropriate pH gradient for the eluting buffer the various components which have initially bound to the fixed ions can be eluted in sequence from the column.

The use of iso-electric focusing enables the iso-electric pH (p*I*) of ampholytes such as proteins to be determined giving information regarding the effect of pH on their ionization. Refer to 'Ionic Separations' (p. 73) for background information and also to *Analytical Biochemistry* (§ 3.3.1, 3.3.3, 11.3.6).

ANALYTICAL PROBLEM

A tissue extract contains a mixture of proteins and it is necessary to separate and purify the components, as far as possible, using ion-exchange chromatography. The appropriate ion-exchange medium and analytical conditions need to be selected.

The following questions need to be answered:

A Is the most appropriate ion-exchange medium an anion or a cation exchanger?

B Which pH conditions give optimum separation?

INVESTIGATIONS

The sample was studied using an iso-electric focusing technique.

The sample was separated by ion-exchange chromatography using various ion-exchange media. The pH range was selected on the basis of the results obtained from the iso-electric focusing investigation.

DATA

The separation pattern for the sample using iso-electric focusing is shown in Figure 16a.

The chromatograms obtained by ion-exchange column chromatography over various pH ranges using two different ion-exchange media are represented in Figure 16b.

Type of medium	*Initial pH selected*	*Range and direction of pH*	*Chromatogram number*
CAX*	7	7–4	1
CAX*	4	4–7	2
CAX*	6	6–4	3
CCX†	7	7–4	4
CCX†	4	4–7	5
CCX†	6	6–4	6

* Anion exchanger, cellulose based, CAX. Working pH range 2–7.

† Cation exchanger, cellulose based, CCX. Working pH range 6–10.

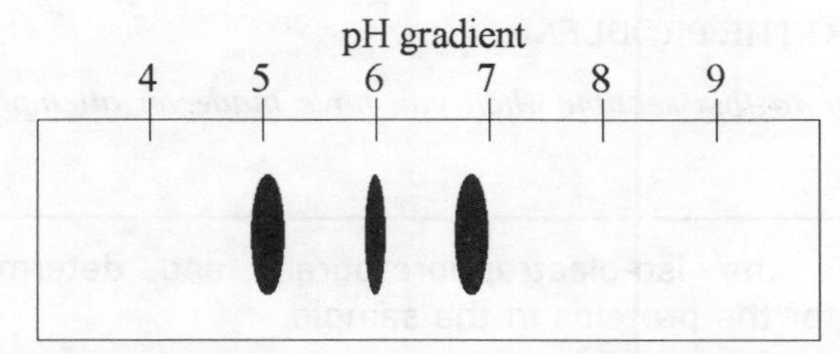

Fig. 16 (a) Iso-electric focusing pattern of the tissue extract.

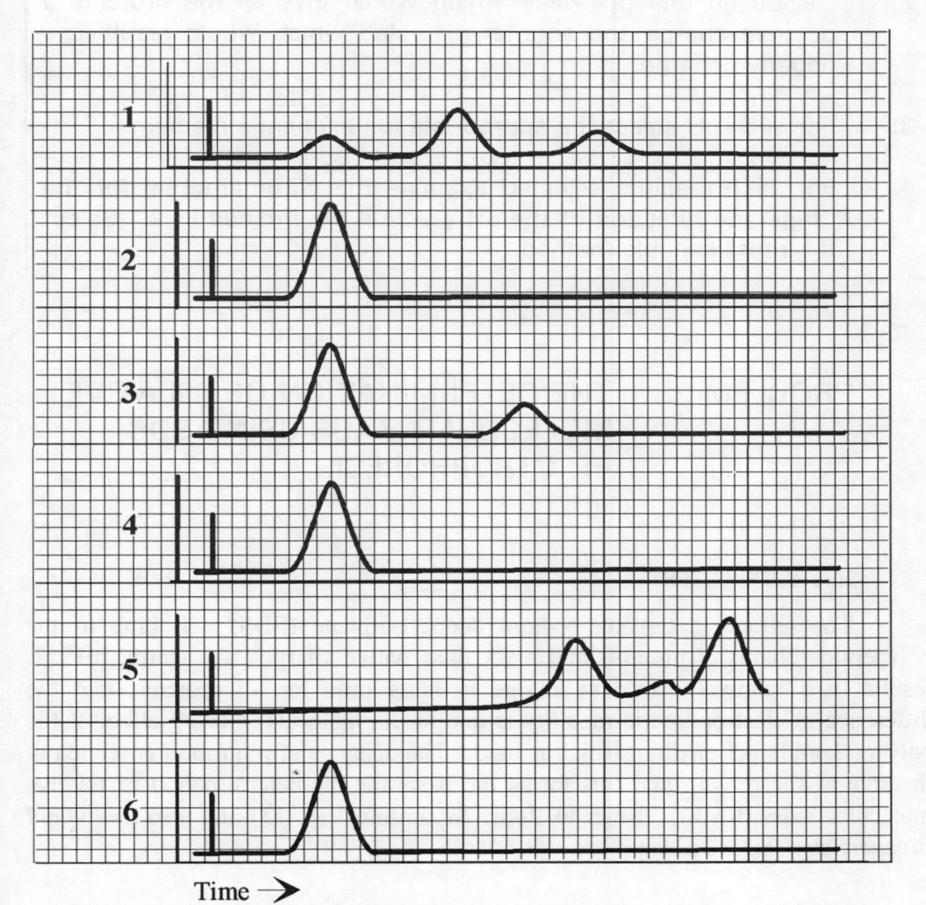

(b) Ion-exchange separations using different media, showing a representation of six chromatograms produced as specified in the text. The effluent from the column was monitored spectrophotometrically at 280 nm for the presence of protein.

SOLVING THE PROBLEM

Only refer to this section when you have made an attempt to answer the question.

1. Examine the iso-electrophoretogram and determine the p*I* values for the proteins in the sample.

2. Decide on one pH value which would give all the proteins a positive charge and another pH value which would result in a negative charge.

3. For each pH select the appropriate ion-exchange medium.

4. For each medium selected examine the chromatogram for the range and direction of the pH gradient determined in 2. Select the most suitable method.

PROBLEM 25 THE DETERMINATION OF RELATIVE MOLECULAR MASS USING SDS ELECTROPHORESIS

INTRODUCTION

The anionic detergent, sodium dodecylsulphate (SDS) will bind to the polypeptide chains of proteins and mask their native charge. At neutral pH this results in a relatively constant charge to mass ratio for all proteins, and the difference in electrophoretic mobility is due to the molecular-sieving effect of the polyacrylamide gel medium which is used. The distance of migration is related to the size of the protein and this forms the basis for the determination of relative molecular mass (RMM). Refer to 'Ionic Separations' (p. 73) and also *Analytical Biochemistry* (§ 3.3.2, 11.3.3).

ANALYTICAL PROBLEM

In an investigation of the properties of a protein (X) it was subjected to the action of a hydrolytic enzyme. The size of any digestion products would give information useful in determining the primary structure of the protein.

The following questions need to be answered:

A Has the enzymatic treatment effected fragmentation of the protein?
B What is the RMM of the original protein and any products of hydrolysis?

INVESTIGATIONS

The original protein (X), the enzyme treated protein (X) and markers of known RMM were prepared for SDS electrophoresis by heating at 100 °C for 2 min in a buffer solution containing SDS and β-mercaptoethanol. After electrophoresis the gel was stained with Coomassie blue to visualize the bands.

DATA

The electrophoretogram is shown in Figure 17.

Lane		*RMM*
1	Markers	14 400
		20 100
		30 000
		43 000
		67 000
		94 000
2	Protein X	
3	Enzyme	
4	Enzyme–treated protein X	

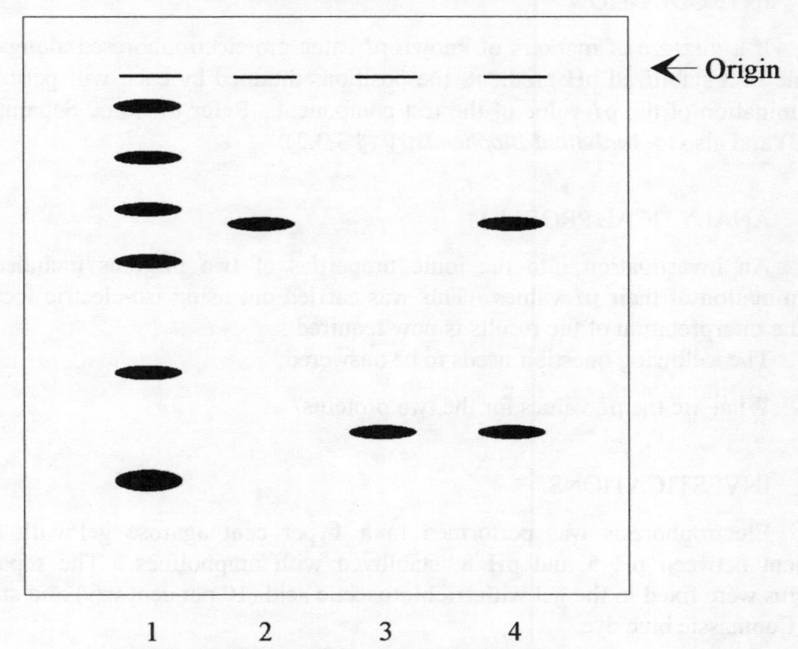

Fig. 17 SDS electrophoresis of protein X.

SOLVING THE PROBLEM

Only refer to this section when you have made an attempt to answer the question.

1. For each of the markers, plot $\log_{10}$ RMM against the distance from the point of application.

2. Measure the distance moved by the samples and determine the RMM of each from the graph.

PROBLEM 26 THE DETERMINATION OF p*I* VALUES USING ISO-ELECTRIC FOCUSING

INTRODUCTION

If a mixture of markers of known p*I* value are electrophoresed alongside a sample in a stabilized pH gradient, the positions attained by each will permit the determination of the p*I* value of the test component. Refer to 'Ionic Separations' (p. 73) and also to *Analytical Biochemistry* (§ 3.3.3).

ANALYTICAL PROBLEM

An investigation into the ionic properties of two proteins included the determination of their p*I* values. This was carried out using iso-electric focusing and the interpretation of the results is now required.

The following question needs to be answered:

- What are the p*I* values for the two proteins?

INVESTIGATIONS

Electrophoresis was performed in a 1 per cent agarose gel with a pH gradient between pH 5 and pH 8, stabilized with ampholines. The separated proteins were fixed in the gel with trichloroacetic acid (10 per cent w/v) and stained with Coomassie blue dye.

DATA

Protein markers

Protein	p*I*
Lentil lectin–acidic	8.15
Horse myoglobin–basic	7.35
Horse myoglobin–acidic	6.85
Human carbonic anhydrase B	6.55
Bovine carbonic anhydrase B	5.85
β–lactoglobulin A	5.20

The electrophoretogram is shown in Figure 18.

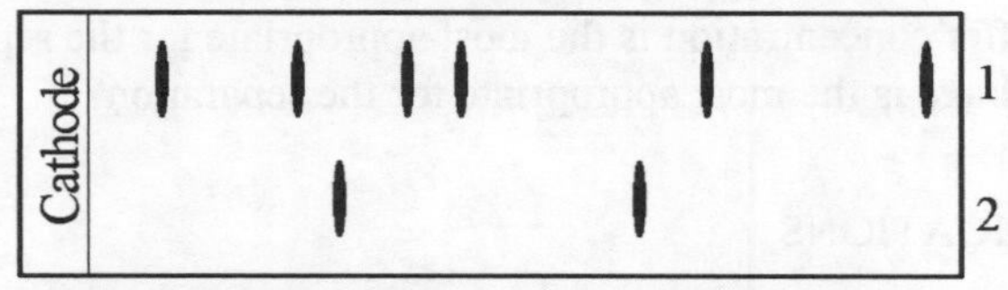

Fig. 18 Determination of p*I* values by iso-electric focusing. Lane 1 contains the two test proteins and lane 2 the marker proteins indicated in the question.

SOLVING THE PROBLEM

Only refer to this section when you have made an attempt to answer the question.

1. Plot the distance moved by the markers (measuring from the cathode end) against their p*I* values.
2. Measure the distance moved by the unknown proteins in a similar manner and read off their p*I* values from the graph.

PROBLEM 27 SELECTION OF SEPARATION CONDITIONS FOR ELECTRO-PHORESIS *(TEST QUESTION)*

INTRODUCTION

Electrophoretic separation depends primarily upon the charge carried by the test substance, and this can be affected by a variety of factors. In developing any such separation method it is essential that these various factors are investigated to enable optimum conditions for the separation to be established. Other factors, external to the charge carried by the test substance, such as electroendosmosis and the structure of the supporting medium, also influence the separation process. Refer to 'Ionic Separations' (p. 73) and also to *Analytical Biochemistry* (§ 3.3.2).

ANALYTICAL PROBLEM

An electrophoretic method is to be developed for the separation of a protein using cellulose acetate membrane as the supporting medium.

The following questions need to be answered:

A What pH is the most appropriate for the separation?
B What buffer concentration is the most appropriate for the separation?
C What voltage is the most appropriate for the separation?

INVESTIGATIONS

Electrophoresis was carried out on a mixture of glucose and the protein at 200 V (20 V cm^{-1} length) for two hours using buffers with a range of pH values and the resulting bands were stained.

Electrophoresis was carried out for one hour on the protein using different voltages and buffers of varying concentrations, but with pH values of 4.0 and 8.0.

DATA

A '+' sign indicates movement towards the anode and a '–' sign indicates movement towards the cathode.

The effect of pH on the migration of the protein and glucose

	Movement (mm)	
pH	Glucose	Protein
4.0	–6	–14
5.0	–6	–2
6.0	–6	+8
7.0	–7	+20
8.0	–8	+32
9.0	–10	+39

The effect of buffer concentration on the migration of the protein

Buffer concentration ($mol\ l^{-1}$)	*Migration* (mm) pH 4.0	pH 8.0
0.025	−35	+78
0.050	−29	+68
0.075	−22	+56
0.100	−12	+45

The effect of voltage on the migration of the protein

Voltage (V)	*Migration* (mm) pH 4.0	pH 8.0
50	−8	+17
100	−14	+37
150	−24	+51
200	−30	+69
250	−39	+89

PROBLEM 28 THE DETERMINATION OF SIZE OF NUCLEIC ACID FRAGMENTS USING AGAROSE ELECTROPHORESIS *(TEST QUESTION)*

INTRODUCTION

Agarose gel electrophoresis can be used to separate molecules on the basis of their size, with the smallest molecules moving the furthest. If a suitable concentration of agarose is selected and markers of known DNA size are electrophoresed alongside, the method can be used to determine the size of the DNA fragments produced as a result of the specific hydrolysis by restriction enzymes. Ethidium bromide is used to visualize the bands after electrophoresis. Refer to 'Ionic Separations' (p. 73) and also to *Analytical Biochemistry* (§ 3.3.2, 13.2.3).

ANALYTICAL PROBLEM

A single fragment of DNA has been isolated from a bacterium. It is required to know whether it contains a recognition site for the restriction endonuclease, BamH1, and also the size of any digestion products.

The following questions need to be answered:

A Does the digest contain more than one fragment?
B What are the sizes of any DNA fragments produced?

INVESTIGATIONS

The sample of DNA was incubated with the enzyme, BamH1.

Agarose gel electrophoresis, using 6.0 g l^{-1} agarose containing 0.5 μg ml^{-1} ethidium bromide, was performed on the digest and a set of DNA markers. The gel was illuminated by ultraviolet light to visualize the bands and a photograph was taken.

DATA

A reproduction of the electrophoretogram is shown in Figure 19. The sizes, in base pairs, of the DNA markers are: 23 130, 9 416, 6 557, 4 361, 2 320, 1 027, 564.

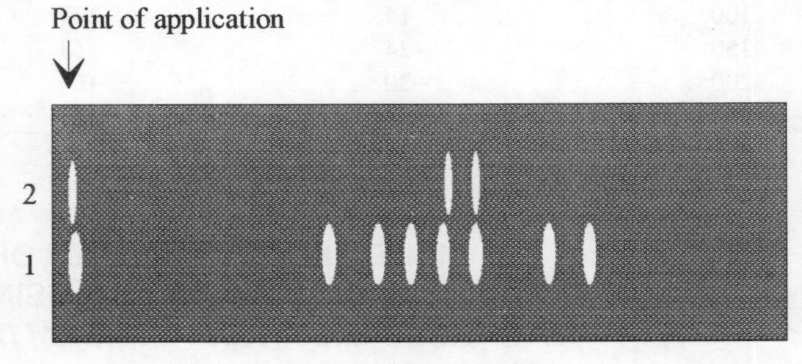

Fig. 19 Electrophoretogram of DNA fragments. Lane 1 contains markers of the sizes indicated in the question. Lane 2 contains the digested sample.

7
RADIOISOTOPES

The atomic nucleus is made up of protons and neutrons. The number of protons determines the atomic number, and hence the identity of an element, and is equal to the number of orbital electrons. The atomic mass is made up by the additional neutrons present. Hence:

atomic number = number of protons
atomic mass = the sum of the number of protons and neutrons

This information is shown as a superscript (atomic mass) and a subscript (atomic number) to the symbol of the element, e.g. $^{125}_{53}I$. In practice the subscript is often omitted because the atomic number is unique to the element which is represented by the appropriate letter, e.g. C for carbon, and for simplicity isotopes are often referred to as carbon-14, phosphorus-32, etc.

The stability of the atomic nucleus depends upon the balance between the repulsive and attractive forces, involving protons and neutrons. Radioactivity is a result of an unstable nucleus and the neutron to proton ratio is an indication of this; elements with N:P ratios of 1.5, or greater, tend to be radioactive.

There are three types of radiation: alpha particles, beta radiation and gamma rays. Each have different properties and associated hazards.

RADIOACTIVE DECAY

Radioactive decay is a random process and, as a result, all measurements must involve statistical assessment. The rate at which a quantity of an isotope decays is proportional to the number of unstable atoms present in the sample, and a graph of activity against time results in a typical exponential curve. It is because of this that the life-span of a radioactive sample cannot be measured, and such decay is normally assessed as the half-life, $t_{1/2}$.

The half-life is defined as the time taken for the activity of a sample to fall by half and is independent of the initial activity. It can be determined using the equation which describes the rate of radioactive decay:

$$\log_e \frac{N_t}{N_0} = -\lambda t$$

where N_t is the activity at time t,
N_0 is the activity at zero time,
λ is the radioactive decay constant,
t equals time.

Converting from natural logarithms the equation becomes:

$$\log_{10} N_t = \log_{10} N_0 - 0.4343\lambda t$$

A graph of the logarithm of the activity of the sample ($\log_{10} N_t$) against time will result in a linear plot and the half-life can be determined from the graph either by measuring the time interval between two readings of activity which vary by a factor of two, or by using the value for the gradient of the line and substituting in the equation. Thus:

$$\text{Half-life } (t_{1/2}) = -\frac{0.3010}{\text{gradient}}$$

The *specific activity* of an isotope indicates the relative amount of radioactive atoms in the sample and is quoted as activity per unit mass or volume, e.g. becquerel per gram.

ISOTOPE DILUTION ANALYSIS

If a known amount of an isotope, with a known specific activity, is mixed with an unknown amount of the non-radioactive substance, the reduction in specific activity can be used to determine the degree of dilution of the isotope and, hence, the amount of non-radioactive substance present. The mixing of the tracer and the test substance must be complete, and it is essential that a pure sample of the substance is subsequently extracted for the determination of its specific activity.

Quantitation is based on the following equation:

$$M_x = M\left(\frac{S}{S_x} - 1\right)$$

where M_x is the amount of unknown substance,
M is the amount of isotope,
S is the specific activity of the isotope,
and S_x is the specific activity of the mixture.

PROBLEM 29 THE USE OF A RADIOACTIVE TRACER TO ASSESS THE EFFICIENCY OF AN EXTRACTION PROCEDURE

INTRODUCTION

Radioactive atoms can be used to label compounds and will enable their presence to be monitored with a considerable degree of sensitivity. The fate of the labelled compound can be assumed to be true for the identical, but unlabelled, compound. Used in this manner, such labelled compounds are known as 'tracers'. Refer to 'Radioisotopes' (p. 87) and also to *Analytical Biochemistry* (§ 5.3.1).

ANALYTICAL PROBLEM

In the development of an assay for thyroxine, blood plasma was required which contained no thyroxine. This was prepared by removing the hormone by adsorption on powdered charcoal. It was necessary to know whether the extraction process was satisfactory, and this was assessed using labelled thyroxine.

The following question needs to be answered:

■ How efficient is the extraction process?

INVESTIGATIONS

Thyroxine labelled with ^{125}I was added to blood plasma and replicate aliquots were removed for gamma counting (sample A).

1 g of powdered charcoal was added to 10 ml of plasma and the mixture stirred for 16 hours at 4 ^{O}C. The resulting slurry was centrifuged at 20 000 × g. Replicate aliquots of the supernatant fluid were removed for gamma counting (sample B).

DATA

Sample	*Counts per minute* (mean of replicates)
A (untreated)	18 135
B (charcoal treated)	424
Background count	130

SOLVING THE PROBLEM

Only refer to this section when you have made an attempt to answer the question.

1.	Calculate the activity remaining after charcoal extraction as a percentage of the untreated sample, taking the background count into consideration.

PROBLEM 30 THE USE OF ISOTOPE DILUTION ANALYSIS FOR QUANTITATION

INTRODUCTION

Isotope dilution analysis is appropriate when there is available both a method for the complete separation of the compound and also a radioactively labelled preparation of that compound of known specific activity. The addition of an unknown amount of the test compound, which is not radioactive, to a known amount of the labelled compound, will not affect the total radioactivity of the mixture, but will reduce the specific activity. This forms the basis of a quantitative method for the test compound. Refer to 'Radioisotopes' (p. 87) and also to *Analytical Biochemistry* (§ 5.3.2).

ANALYTICAL PROBLEM

Vitamin B_{12} can be quantitated by isotope dilution analysis. This assay involves the use of ^{57}Co-B_{12} and is based on the fact that vitamin B_{12} is bound specifically by intrinsic factor and the free vitamin can be adsorbed by protein-coated charcoal and removed by centrifugation. The vitamin which is complexed to intrinsic factor is not adsorbed by the charcoal. Data is available from a vitamin B_{12} assay.

The following question needs to be answered:

■ What is the concentration of vitamin B_{12} in the sample (pg ml^{-1})?

INVESTIGATIONS

The assay was performed as follows:

1. 1 ml of a solution of ^{57}Co-B_{12} (1000 pg ml^{-1}) was added to 1 ml of a sample and to two control tubes, intrinsic factor control (IFC) and adsorption control (AC).
2. Intrinsic factor was added to the test and IFC tubes but not to the AC tube

and allowed to react. The final volume of all tubes was the same.

3. The remaining free vitamin (both labelled and unlabelled) was removed by adding coated charcoal to all tubes and centrifuging after 30 minutes incubation.
4. The radioactivity in the supernatant (complexed vitamin) was counted in all tubes using a gamma radiation counter.

DATA

Sample	*Added* $^{57}Co\text{-}B_{12}$ (pg)	*Counts per minute* *
Test	1000	1340
IFC	1000	2450
AC (no intrinsic factor)	1000	15

* Each value is the mean of three measurements.

SOLVING THE PROBLEM

Only refer to this section when you have made an attempt to answer the question.

1. Refer to 'Radioisotopes' (p. 87) and substitute into the equation given for isotope dilution analysis. Bear in mind that because the volumes are identical the actual count is effectively the specific activity.

2. Allow for the background count which, if adsorption is effective, should be minimal.

PROBLEM 31 THE CALCULATION OF HALF-LIFE OF A RADIOISOTOPE *(TEST QUESTION)*

INTRODUCTION

The fact that the activity of a radioisotope is decreasing all the time has significant implications from an analytical point of view. If an analytical method depends upon radioactivity, the reagents have a limited life-span and as time progresses the sensitivity of the method decreases. It is essential to be able to calculate the half-life of an isotope in order to be able to assess the effect of time upon the activity of the reagents. Refer to 'Radioisotopes' (p. 87) and also to *Analytical Biochemistry* (§ 5.1.2).

ANALYTICAL PROBLEM

A sample of a particular isotope is required for analytical purposes with a minimum activity of 5000 counts per minute (cpm). In organizing the delivery of the isotope to the laboratory it is necessary to know the half-life in order to be able to specify the activity required when the sample is dispatched. The time taken to transport the isotope from the supplier is 4½ hours.

The following questions need to be answered:

A What is the half-life of the isotope?

B What activity of isotope must be purchased from the supplier in order that it will have a minimum activity of 5 000 cpm on arrival at the laboratory?

INVESTIGATIONS

The activity of a sample of the isotope was monitored over a period of time.

DATA

Decay of activity in the sample

Time (min)	0	60	120	180	240	300	360
Activity (disintegration min^{-1})	248	224	202	182	164	148	134

PROBLEM 32 THE USE OF ISOTOPE DILUTION ANALYSIS FOR DETERMINING BLOOD VOLUME *(TEST QUESTION)*

INTRODUCTION

Chromium ions are readily taken up by red blood cells with no impairment of their function. This phenomenon can be used in an isotope dilution technique for the measurement of blood volume if radioactive chromium is used as the tracer. Refer to 'Radioisotopes' (p. 87) and also to *Analytical Biochemistry* (§ 5.3.2).

ANALYTICAL PROBLEM

In the investigation of physiological function it was necessary to determine the total blood volume of an animal. The method of isotope dilution analysis was undertaken and the data obtained.

The following question needs to be answered:

- What is the blood volume as determined by isotope dilution analysis?

INVESTIGATIONS

A 10 ml suspension of ^{51}Cr labelled red blood cells was injected into the animal. After 10 minutes a small blood sample was taken and the radioactivity measured.

DATA

Sample	*Radioactivity* (cpm ml^{-1})
Labelled cells	3×10^7
Mixed sample	5×10^4

8

ENZYME ASSAYS

Enzymes are proteins which catalyse various reactions. They influence the rate of a reaction, i.e. conversion of substrate to products, and their activity is measured in terms of this rate or velocity. This measurement can be done in a variety of ways,

MONITORING TECHNIQUES

The rate of the reaction, in terms of substrate depletion or product accumulation, may be monitored by recording the concentration change and plotting a reaction progress curve. Such a technique is known as a *kinetic or continuous assay*. The most valid measure of the rate of the reaction is the initial velocity, V_0. If this is not calculated by the recording instrument it may be determined by drawing a tangent to the curve at zero time. Hence:

$$\text{Enzyme concentration} \propto \text{velocity } (V_0)$$

Fixed time assays measure the amount of substrate used, or the amount of product formed, over a relatively long but specified period of time. Hence:

$$\text{Enzyme concentration} \propto \frac{1}{\text{amount of substrate used}}$$

$$\text{Enzyme concentration} \propto \text{amount of product formed}$$

A technique which depends upon the same basic assumptions as do fixed time assays, but relates the enzyme activity to the time taken for a fixed amount of product to be formed, is called a *fixed change assay*. It is particularly useful for reactions which result in a pH change and are monitored potentiometrically. Hence:

$$\text{Enzyme concentration} \propto \frac{1}{\text{time}}$$

OPTIMIZATION OF ASSAYS

The catalytic activity of enzymes is dependent on the maintenance of their native structure, and any slight variations may result in significant changes in this activity. Thus enzyme assays must be performed under controlled conditions which are optimal in terms of pH, presence of activators or inhibitors, substrate type and concentration. The selection of assay temperature may also be important.

Enzymes are very sensitive to changes in pH and function best over a limited range with a definite pH optimum. This optimum will often vary from one substrate to another and must be determined for each substrate. The buffer system used will often affect the overall activity of an enzyme and may alter its pH optimum.

When excess substrate is used, an enzyme reaction shows maximum velocity and this is proportional to the amount of enzyme present. In order to achieve this, a substrate concentration which is at least ten times greater than the Michaelis constant (K_m) for the enzyme should be used. While it is desirable to use as high a concentration as possible, some enzymes are subject to inhibition at very high substrate levels and in such cases the chosen concentration must give the highest reaction rate possible.

The concept of the formation and dissociation of an enzyme-substrate (ES) complex during an enzyme reaction was developed by Michaelis and Menten who derived an equation which is crucial to enzyme studies.

$$v = \frac{V_{max} \times [S]}{K_m + [S]}$$

The equation gives a measure of the K_m, the dissociation constant for the ES complex, in terms of the measured velocity of the reaction (v) which results from a substrate concentration (S) and the maximum velocity (V_{max}) which can be achieved using a very high concentration of substrate.

The value for maximum velocity is related to the amount of enzyme used, but the K_m is peculiar to the enzyme and is a measure of the efficiency of the enzyme as a catalyst. Enzymes with large values for K_m show a low affinity for the substrate and hence are often less active than enzymes with a low K_m value.

Determination of the K_m value involves plotting the reaction velocities, produced by a fixed amount of the enzyme, with different concentrations of substrate. The substrate concentration that results in a velocity which is half the maximum velocity is numerically equal to K_m.

While this method is extremely simple, it is also experimentally inaccurate because of the difficulty in determining the maximum velocity accurately.

Lineweaver and Burk described a method for the determination of K_m which uses the reciprocal form of the Michaelis equation converting it to a linear relationship.

$$\frac{1}{v} = \frac{K_m}{V_{max}} \times \frac{1}{[S]} + \frac{1}{V_{max}}$$

A plot of the reciprocal of the velocity against the reciprocal of the substrate concentration will give a straight line graph with intercepts of:

$$\frac{1}{[S]} = -\frac{1}{K_m}$$

and

$$\frac{1}{v} = \frac{1}{V_{max}}$$

It is essential when implementing an enzyme assay, to determine the effective range of activity for which the relationship between reaction rate and enzyme activity is linear. This is most conveniently done by assaying carefully prepared dilutions of a sample containing a high enzyme activity and plotting the resulting reaction rate against the percentage dilution of the sample.

In *coupled assays*, the reaction mixture includes the substrates for the initial or test enzyme, and also the additional enzymes and reagents necessary to convert the product of the first reaction into a detectable product for the final measured reaction. In some instances there may be some degree of inhibition between the different enzyme reactions and these must be investigated. In such complex assays it is often difficult to decide on assay conditions. In designing an enzyme assay the main consideration is to select conditions which give maximum reaction rate for the test enzyme to ensure maximum sensitivity for the method.

UNITS OF ENZYME ACTIVITY

A *katal* is defined as that amount of enzyme which will result in the conversion of one mole of substrate to product in one second. A convenient sub-unit is the nanokatal.

An older unit of activity, which is still used, is the *International Unit* which is defined as that amount of enzyme which will result in the conversion of one micromole of substrate to product in one minute.

1.0 nanokatal = 0.06 International Units

The *specific activity* of an enzyme preparation is expressed as the units of activity per mg of protein and is a convenient way of comparing the purity of enzyme preparations.

CALCULATION OF ENZYME ACTIVITY

Considerable difficulty is often experienced in calculating the activity from first principles and it is important to appreciate the necessary stages in the calculation.

1. The concentration of the product formed in one second, expressed as moles per litre, is calculated. For fixed time assays this most frequently involves the use of known concentrations of the product and a calibration graph. It is most convenient if molar absorption coefficients can be used in spectrophotometric assays with the elimination of the need for standards.
2. The actual amount of product formed in one second in the assay volume used is calculated, and this is equal to the number of units of enzyme activity in the assay.
3. The volume of the sample used in the assay must be accounted for and the activity quoted for a standard volume of sample, e.g. katals per millilitre.

PROBLEM 33 THE OPTIMIZATION OF ASSAY CONDITIONS OF A COUPLED KINETIC ENZYME ASSAY

INTRODUCTION

The measurement of initial velocity provides the most acceptable data for enzyme activity measurements but the assay conditions must be optimal for the test enzyme in order to offer maximum sensitivity. In designing an enzyme assay it is essential to first determine the optimal conditions for all the enzymes involved, and then decide on the assay conditions before determining the concentration of substrate necessary to give maximum enzyme activity. Refer to 'Enzyme Assays' (p. 94) and also to *Analytical Biochemistry* (§ 8.1.1, 8.1.2, 8.2.2, 8.3.1, 8.3.2).

ANALYTICAL PROBLEM

It is intended to develop a kinetic assay for the enzyme R-amine oxidase (RAO) using the following coupled reaction involving glutamate dehydrogenase (GDH) as the indicator enzyme:

$$\text{R-amine} + O_2 \xrightarrow{\text{RAO}} R + H_2O_2 + NH_3$$

$$NH_3 + \text{oxoglutarate} + \text{NADH} \xrightarrow{\text{GDH}} \text{glutamate} + \text{NAD}$$

The following questions need to be answered:

A What pH would be selected for the assay?
B What is the optimal concentration for the substrate, R-amine?
C What co-factors are required and at what concentrations?

INVESTIGATIONS

A series of experiments were undertaken to study:

- the effect of pH on the activity of each enzyme separately (Figure 20),
- the effect of the co-factors flavin adenine dinucleotide (FAD) and adenosine diphosphate (ADP) on the activity of the two enzymes,
- the effect of R-amine concentration on the rate of the coupled reaction.

DATA

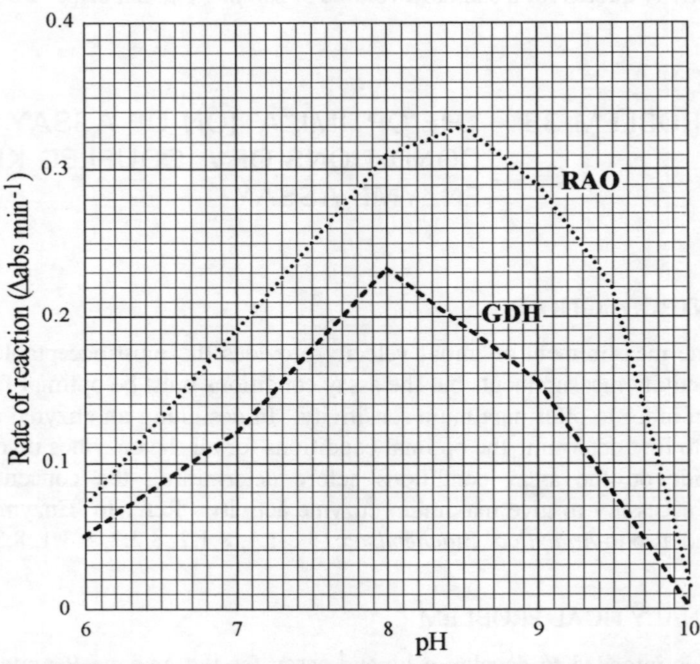

Fig. 20 The effect of pH on the activity of R-amine oxidase (RAO) and glutamate dehydrogenase (GDH).

Effect of mixtures of concentrations of FAD and ADP on the activities of the enzymes RAO and GDH as measured by reaction rate

FAD Concentration (mmol l^{-1})	*Reaction rate* (Abs min^{-1}) — *ADP Concentration* (mmol l^{-1}): None		0.01		0.1		1.0	
	RAO	GDH	RAO	GDH	RAO	GDH	RAO	GDH
None	0.3	0.3	0.2	0.4	0.1	0.9	0.01	1.4
0.01	1.0	0.3	1.0	0.4	0.6	1.0	0.5	1.4
0.1	1.1	0.3	1.0	0.4	1.1	1.0	1.1	1.4
1.0	1.1	0.3	1.1	0.4	1.1	1.0	1.1	1.4

Effect of substrate concentration of the activity of the enzyme RAO

Substrate concentration (μmol l^{-1})	*Reaction rate* (Abs min^{-1})
20.0	0.238
25.0	0.270
38.5	0.376
77.0	0.589
111.0	0.715
200.0	0.952
400.0	1.176

SOLVING THE PROBLEM

Only refer to this section when you have made an attempt to answer the question.

1. From the graph, determine the optimum pH for each enzyme.
2. Determine the K_m value for R-amine oxidase by the Lineweaver-Burk method. Using this information, specify the working concentration for the substrate.
3. Deduce any activating or inhibiting effects of the co-enzymes.
4. Determine the optimal concentrations for ADP and FAD for the assay.

PROBLEM 34 THE DEVELOPMENT OF A FIXED TIME ENZYME ASSAY

INTRODUCTION

Many enzyme assay methods involve measuring the total reaction which takes place in a fixed reaction time rather than measuring the initial reaction velocity. Such methods are usually relatively simple and many are photometric in nature. In designing a method it is important that conditions are selected such that there is a definite relationship between the readings obtained and the amount of enzyme present. Refer to 'Enzyme Assays' (p. 94) and also to *Analytical Biochemistry* (§ 2.2.3, 8.3.1, 8.3.2).

ANALYTICAL PROBLEM

A method for the assay of an enzyme has been proposed in which the amount of product formed in the reaction can be detected colorimetrically using a specific reagent. A reaction time needs to be selected which will give valid results.

The following questions need to be answered:

A What is a potentially suitable reaction time?

B Is there a linear relationship between the measurements taken and the amount of enzyme present using the proposed reaction time?

INVESTIGATIONS

An assay reaction was set up in bulk. Samples were removed at regular time intervals and the amount of product measured by the addition of a colour reagent.

Using the proposed reaction time, a series of dilutions of an enzyme sample were assayed.

NB. For simplicity, the data for only a limited selection of assay times is provided. Use the data given for a reaction nearest to your proposed reaction time.

DATA

The absorbance of the product of the reaction was measured in aliquots of the reaction mixture at 2 min intervals during the reaction.

Time (min)	*Absorbance*
0	0
2	0.36
4	0.60
6	0.84
8	1.05
10	1.18
12	1.26
14	1.35
16	1.39
18	1.40
20	1.40

A series of dilutions of an enzyme preparation were assayed using assay times of 2, 5, 10 and 18 min.

Enzyme Dilution (%)	*Absorbance due to product concentration (mean of three reading)* Reaction Time (min) 2	5	10	18
10	0.04	0.08	0.16	0.28
20	0.08	0.16	0.31	0.53
30	0.12	0.22	0.47	0.74
40	0.14	0.30	0.61	0.94
50	0.16	0.39	0.78	1.12
60	0.20	0.46	0.91	1.22
70	0.22	0.53	1.05	1.27
80	0.27	0.61	1.15	1.31
90	0.29	0.69	1.25	1.31
100	0.33	0.76	1.30	1.34

SOLVING THE PROBLEM

Only refer to this section when you have made an attempt to answer the question.

1. Draw a graph of the reaction progress from the first set of data and select the most appropriate reaction time for the assay.

2. Plot the data for each of the four fixed time assays. Over what range is the assay valid for each reaction time?

PROBLEM 35 THE CALCULATION OF ENZYME ACTIVITY

INTRODUCTION

It is essential to be able to calculate the activity of the enzyme in katals from the data which is generated from an enzyme assay procedure. Refer to 'Molecular Spectroscopy' (p. 12), 'Enzyme Assays' (p. 94) and also to *Analytical Biochemistry* (§ 8.3.2).

ANALYTICAL PROBLEM

Two assay methods for enzyme X have been developed, one being a kinetic assay and the other a fixed time assay. In both methods the products of the reaction are measured spectrophotometrically at 560 nm. It is now necessary to be able to convert absorbance measurements into enzyme activity per millilitre of sample.

The following question must be answered:

- What is the enzyme activity (katal) per millilitre of sample determined by each method?

INVESTIGATIONS

Kinetic Assay

2.7 ml buffer
0.1 ml co-factor
0.1 ml substrate

The reaction was started by adding 0.1 ml enzyme X preparation. The absorbance was monitored at 560 nm using a recording spectrophotometer.

Fixed time assay

4.7 ml buffer
0.1 ml co-factor
0.1 ml substrate

The reaction was started by adding 0.1 ml enzyme X preparation. The mixture was incubated for 10 min and the reaction stopped by adding 5.0 ml colour reagent. The absorbance of the mixture was measured in triplicate at 560 nm.

A standard solution of the product of the reaction containing 0.5 mmol l^{-1} was treated as follows:

5.0 ml standard solution
5.0 ml colour reagent

The absorbance of the mixture was measured in triplicate at 560 nm.

DATA

Kinetic assay

A trace from the recording spectrophotometer set at 560 nm is shown in Figure 21.

The molar absorption coefficient of the product at 560 nm is 5.0×10^3 $l\ mol^{-1}\ cm^{-1}$.

Fixed time assay

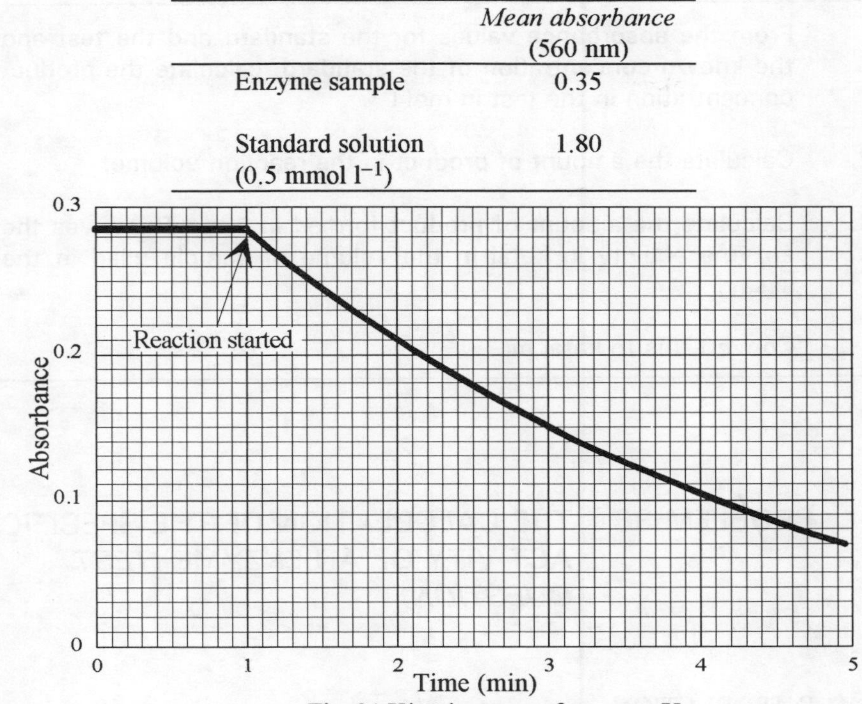

	Mean absorbance (560 nm)
Enzyme sample	0.35
Standard solution (0.5 mmol l^{-1})	1.80

Fig. 21 Kinetic assay of enzyme X.

SOLVING THE PROBLEM

Only refer to this section when you have made an attempt to answer the question.

Kinetic assay

1. Draw a tangent to the reaction trace and determine the initial rate of reaction in absorbance change per second.

2. Using the Beer–Lambert equation (refer to 'Molecular Spectroscopy', p. 12), calculate the concentration of product produced over the same period.

3. Calculate the amount of product produced in the assay volume. This is the enzyme activity present in the sample volume used in the assay.

4. Convert this to katal ml^{-1}.

Fixed time assay

1. From the absorbance values for the standard and the test and the known concentration of the standard, calculate the product concentration in the test in mol l^{-1}.

2. Calculate the amount of product in the reaction volume.

3. Calculate the amount of product formed in 1 s. This gives the enzyme activity in katal in the volume of sample used in the assay.

4. Convert this to katal ml^{-1}.

PROBLEM 36 THE CALCULATION OF THE SPECIFIC ACTIVITY OF AN ENZYME *(TEST QUESTION)*

INTRODUCTION

The production of an enzyme for analytical use normally involves various extraction and purification processes. It is necessary to know the activity and purity of enzymes used in analytical methods. This is usually expressed in terms of the specific activity. Refer to 'Enzyme Assays' (p. 94), 'Molecular Spectroscopy' (p. 12) and also to *Analytical Biochemistry* (§ 8.3.2).

ANALYTICAL PROBLEM

The enzyme R-amine oxidase (RAO) has been extracted from a sample of tissue and it is necessary to assess the purity of the preparation. The sample has been analysed by the method outlined in Problem 33, which involves the oxidation of NADH and monitoring the fall in absorbance at 340 nm. In this instance, however, because the sample is still in a relatively crude form, it shows a blank reaction, i.e. there is some oxidation of NADH before the substrate for the RAO is added. This fall in absorbance is not due to the RAO reaction, and an appropriate correction of the initial rate of the enzyme reaction must be made.

The protein content of the sample has also been determined.

The following questions need to be answered:

A What is the enzyme activity of the enzyme preparation?
B What is the protein content of the enzyme preparation?

C What is the specific activity of the enzyme preparation?

INVESTIGATIONS

The activity of the preparation of the enzyme RAO was measured by adding 0.1 ml of the sample to 2.9 ml of the reagent mixture. The fall in absorbance at 340 nm due to the oxidation of NADH was monitored both before and after the addition of the substrate R-amine.

The protein content was assessed by measurement of absorbance at 280 nm.

DATA

Enzyme assay

The trace for the kinetic assay of the sample is given in Figure 22.

The molar absorbance coefficient for NADH at 340 nm is 6.22×10^3 $l\ mol^{-1}\ cm^{-1}$.

Protein determination

Standard protein ($g\ l^{-1}$)	*Absorbance at 280 nm*
1.0	0.15
2.0	0.24
3.0	0.32
4.0	0.39
5.0	0.44
Sample	0.28

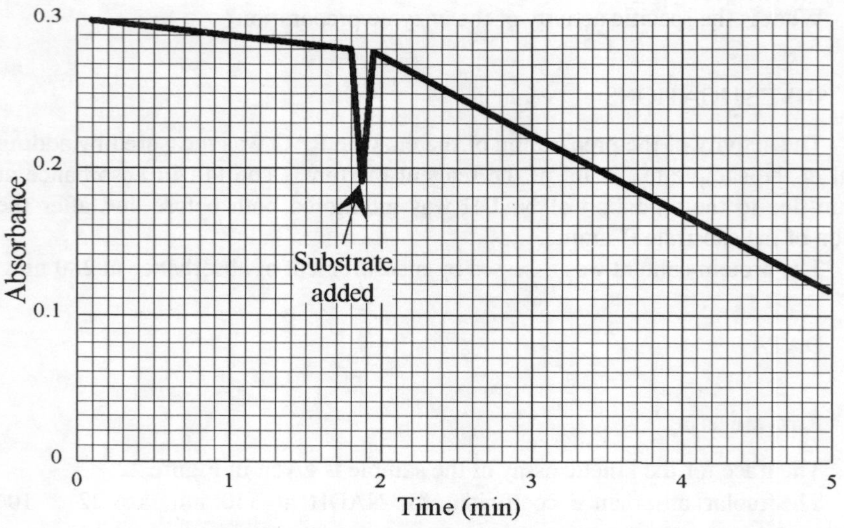

Fig. 22 Kinetic assay of a sample of R-amine oxidase.

PROBLEM 37 COMPARISON OF THE PURITY OF TWO ENZYME PREPARATIONS *(TEST QUESTION)*

INTRODUCTION

There are a variety of ways of purifying crude enzyme preparations, and it is necessary to determine the relative purity of the samples in order to comment on the efficiency of any method. The specific activity is the most acceptable way of expressing the purity of an enzyme preparation. Refer to 'Enzyme Assays' (p. 94) and also to '*Analytical Biochemistry*' (§ 8.3.2).

ANALYTICAL PROBLEM

The enzyme alkaline phosphatase was extracted and purified using two different procedures (A and B) as a pilot scheme before deciding on a method for routine use.

The following questions need to be answered:

A What is the specific activity of each sample?
B Which extraction method yields the purest enzyme preparation?

INVESTIGATIONS

The two samples of the enzyme alkaline phosphatase were assayed using disodium phenylphosphate as a substrate and measuring the amount of phenol liberated by the enzyme, using the reaction with Folin and Ciocalteu reagent to give a blue colour. A calibration graph was prepared from standard solutions of phenol.

The protein content of each sample was determined using the biuret reaction and a series of albumin standards.

DATA

Enzyme assay

2.0 ml buffer solution
2.0 ml substrate solution
0.1 ml enzyme preparation
Incubate for 10 min and measure the absorbance at 600 nm after the addition of the colour reagent.

Phenol standards

2.0 ml buffer solution
2.0 ml phenol standard solution
0.1 ml distilled water
Measure the absorbance at 600 nm after the addition of the colour reagent.

Phenol standards (mmol l^{-1})	*Absorbance at 600 nm*
1.0	1.95
0.8	1.60
0.6	1.20
0.4	0.73
0.2	0.38
Samples	
Enzyme A	0.88
Enzyme B	0.74

Protein determination

1.0 ml sample
4.0 ml biuret reagent
Measure the absorbance at 540 nm after 30 min.

Protein standards ($g\ l^{-1}$)	*Absorbance at 540 nm*
50	1.70
40	1.40
30	1.15
20	0.75
10	0.37
Samples	
Enzyme A	0.18
Enzyme B	0.97

9

AUTOMATED FLOW ANALYSIS

In automated flow analysis samples are analysed as they are pumped in sequence through a series of tubing containing a continuously flowing reagent stream. This tubing connects the different units of the analyser, which each perform a specific analytical function. The simple analysers are mainly based on spectrophotometric methods of analysis and are of modular design, appropriate modules being selected for each particular assay. These include a sampler, heating bath, pump, dialyser and detector.

As distinct from manual methods, absolute volumes are not pipetted, but the liquids involved are mixed in carefully controlled proportions. These ratios are determined by using manifold tubes of varying internal diameter with a constant speed pump, and it is the different rates of flow (ml min^{-1}) within these tubes that is the important factor. Whenever two liquid streams are brought together a glass coil is included in the system to effect adequate mixing. These coils can also be used to prolong the reaction time.

Quantitation is based on the direct comparison of the peak heights on a recorder trace of the samples, and the standards, using either a calibration curve or a computer. It is not necessary for the reaction to go to completion because all the measurements in a particular method are made after the same fixed reaction time, which is determined by the length of analyser tubing and relies on a constant pump speed.

MANIFOLD AND FLOW DIAGRAM

The introduction of the sample and reagents in predetermined proportions is achieved by using flexible tubes of different internal diameters, referred to as manifold tubes. These must be selected for a particular method.

The word manifold refers to the distinctive assembly of tubes, glass mixing coils and interconnecting tubing which is required for a particular assay. Information on the modules required is usually also included in *flow injection analysis* (Figure 23).

A more detailed diagrammatic representation of the components and layout of the system required to perform a particular method is known as a flow diagram. It specifies the sampling rate, manifold tubes, units required, the type of detection system, etc. This is the recognized method of giving the information required in *air-segmented continuous flow analysis* systems (Figure 24).

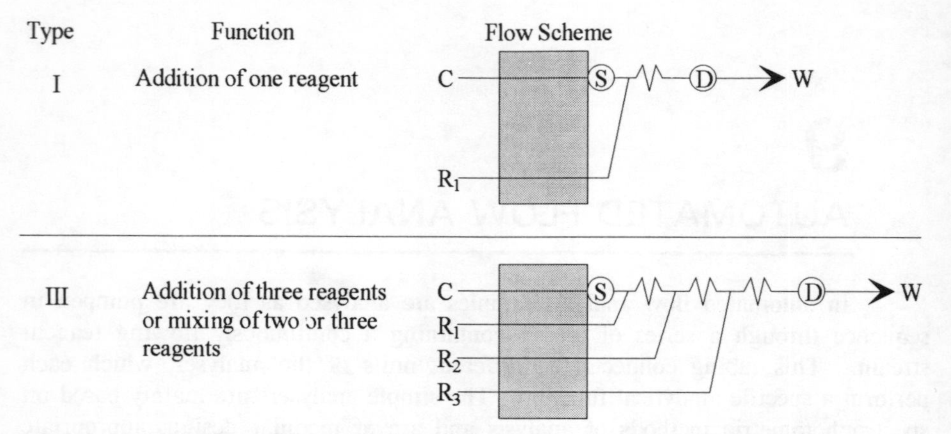

Fig. 23 Manifolds for flow injection analysis. (S) sample injection port, (C) carrier stream, (R_1,R_2,R_3) reagent streams, (D) detector, (W) waste.

CARRY-OVER

The interaction between adjacent samples as they flow through the tubes following each other is known as carry-over. It must be kept to a minimum if reliable results are to be obtained. Two factors which help to reduce carry-over in air-segmented flow analysis are the introduction of air bubbles into the liquid streams and the aspiration of water between successive samples. In flow injection analysis excessive dispersion of the sample zone in the liquid stream is reduced by use of small sample volumes, narrow bore tubing and short residence times in the system.

Carry-over manifests itself on the recorder trace as peaks which are not completely differentiated. This is most noticeable when a sample of low concentration follows one of high concentration and the peak for the low sample is either seen as a shoulder on the high peak or may even be masked completely.

An acceptable way of determining the percentage interaction is to analyse a sample of high concentration (A) in triplicate, giving results a_1, a_2, and a_3, followed by a sample of low concentration (B) in triplicate b_1, b_2, and b_3. The carry-over between a_3 and b_1 is given by the equation:

$$\text{Percentage interaction}\,(k) = \frac{b_1 - b_3}{a_3 - b_3} \times 100\%$$

Where b_3 is assumed to be the 'true' value for sample B since it is preceded by two samples of equal concentration and the effect of carry-over should be negligible. The determination of the percentage interaction should form part of the method assessment procedure.

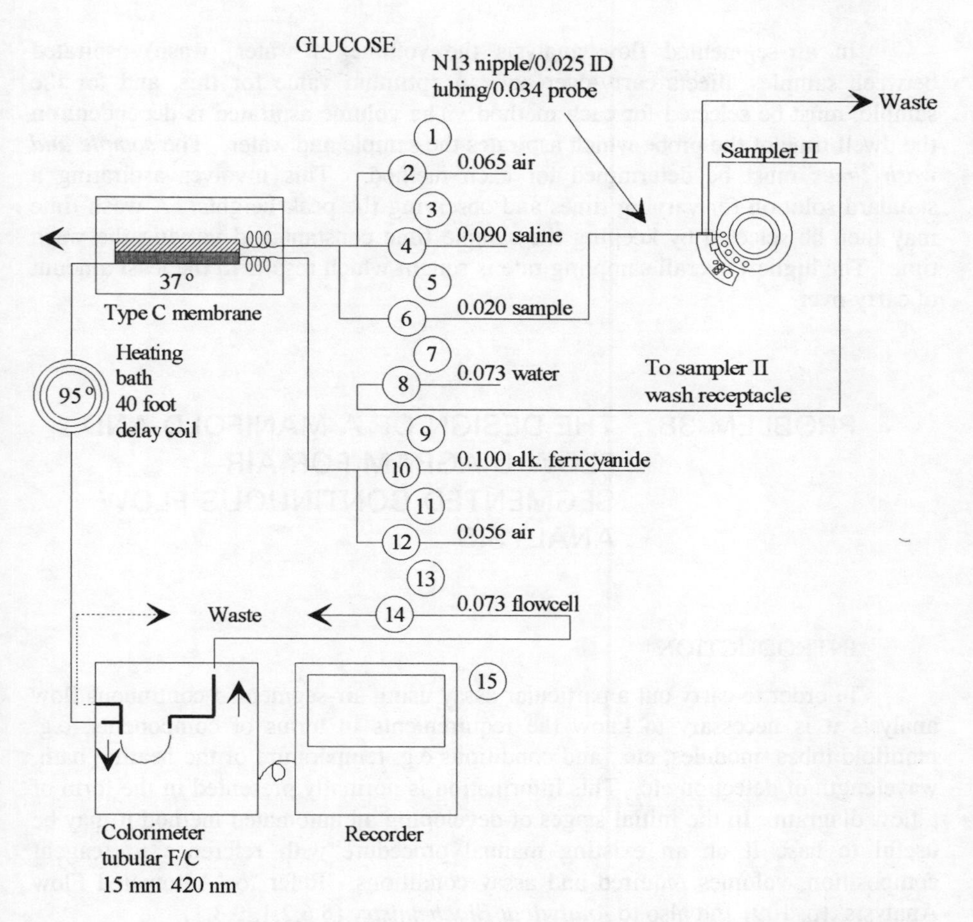

Fig. 24 Flow diagram for the measurement of glucose using alkaline ferricyanide. The pump tubes are specified with respect to their internal diameters and position on the end-blocks. The colour coding of the tubes is:

Position 2	0.065	Blue/Blue
Position 4	0.090	Purple/Black
Position 6	0.020	Orange/Yellow
Position 8	0.073	Green/Green
Position 10	0.100	Purple/Orange
Position 12	0.056	Yellow/Yellow
Position 14	0.073	Green/Green

The odd-numbered positions on the end-blocks are those on the lower level and are unfilled. H3 and D1 refer to the glass connectors.

In air-segmented flow analysis the volume of water (wash) aspirated between samples affects carry-over and an optimum value for this, and for the sample, must be selected for each method. The volume aspirated is dependent on the dwell time of the probe which aspirates the sample and water. The *sample and wash times* must be determined for each method. This involves aspirating a standard solution for varying times and observing the peak heights. A wash time may then be selected by keeping the sample time constant and varying the wash time. The highest overall sampling rate is sought which results in the least amount of carry-over.

PROBLEM 38 THE DESIGN OF A MANIFOLD AND FLOW DIAGRAM FOR AIR-SEGMENTED CONTINUOUS FLOW ANALYSIS

INTRODUCTION

In order to carry out a particular assay using air-segmented continuous flow analysis it is necessary to know the requirements in terms of components, e.g. manifold tubes, modules, etc., and conditions e.g. temperature of the heating bath, wavelength of detection etc. This information is normally presented in the form of a flow diagram. In the initial stages of developing an automated method it may be useful to base it on an existing manual procedure with reference to reagent composition, volumes required and assay conditions. Refer to 'Automated Flow Analysis' (p. 109) and also to *Analytical Biochemistry* (§ 6.2.1, 9.3.1).

ANALYTICAL PROBLEM

It has been suggested that the manual quantitative procedure for aqueous glucose samples using the enzyme, glucose oxidase, could be performed by air-segmented flow analysis using a simple modular instrument. In order to investigate this possibility an experimental flow diagram must be designed based on the manual procedure.

The following questions need to be answered:

A Which modules are required?

B What would be the lay-out of the flow diagram?

INVESTIGATIONS

The manual assay for reagent volumes, assay conditions and steps in the analysis was consulted.

DATA

Manual Procedure

To 4.0 ml glucose oxidase reagent, 0.1 ml sample was added. This was mixed and allowed to react for 30 min at room temperature. The absorbance was measured at 560 nm.

Manifold tubing

Flow rate (ml min^{-1})	*Colour coding*	*Flow rate* (ml min^{-1})	*Colour coding*
0.015	orange/black	1.00	grey
0.03	orange/red	1.20	yellow
0.05	orange/blue	1.40	blue/yellow
0.06	red/blue	1.60	blue
0.10	orange/green	2.00	green
0.16	orange/yellow	2.50	purple
0.23	orange/white	2.90	purple/black
0.32	black	3.40	purple/orange
0.42	orange	3.90	purple/white
0.50	white		
0.80	red		

SOLVING THE PROBLEM

Only refer to this section when you have made an attempt to answer the question.

1. Determine the ratio of volumes used for sample and reagents in the manual procedure.
2. Select the required modules.
3. Draw a proposed flow diagram.
4. Select manifold tubing of appropriate flow rates to give the required volume ratios.

PROBLEM 39 THE ASSESSMENT OF CARRY-OVER IN AIR-SEGMENTED CONTINUOUS FLOW ANALYSIS

INTRODUCTION

Before an assay can be put into routine use on an air-segmented continuous flow analyser it is necessary to determine the sampling rate which should be used, i.e. the number of samples analysed per hour and the ratio of the sample time to wash time, in order to achieve an acceptable carry-over. At high levels of carry-over the interaction between adjacent samples leads to assay imprecision. Some sampler units are designed to achieve a limited selection of sampling rates and sample:wash ratios by the use of a range of cams which fit into the centre of the sampler turntable. Other instruments have variable settings for sample and wash aspiration times. However, in both cases a provisional choice of sampling rate must be assessed for its suitability by determining the percentage carry-over which is produced. Refer to 'Automated Flow Analysis' (p. 109) and also to *Analytical Biochemistry* (§ 6.2.1, 9.3.1).

ANALYTICAL PROBLEM

A method for the measurement of glucose using glucose oxidase has been devised and the sampling rate needs to be determined. The sampler unit is fitted with a cam device and an appropriate cam must be chosen which gives an acceptable level of carry-over.

The following questions need to be answered:

A What is the percentage carry-over using different cams?

B Which cam gives the most acceptable level of carry-over?

INVESTIGATIONS

Carry-over for the system was assessed for two different cams (A and B) by analysing a set of glucose standards, followed by a sample of high glucose content (in triplicate), which was in turn followed by a sample with a low glucose content, also in triplicate.

DATA

The traces for the analyses are shown in Fig. 25 (a and b).

Recorder trace	*Sampling rate* (per hour)	*Sample:wash ratio*
A	40	2:1
B	60	1:1

The triplicate analyses of the low concentration sample are labelled a_1, a_2 and a_3, and the triplicate analyses of the high concentration sample are labelled b_1, b_2 and b_3.

SOLVING THE PROBLEM

Only refer to this section when you have made an attempt to answer the question.

1. For each recorder trace, plot a calibration graph using the set of standards and determine the concentration for peaks a_1, a_2, a_3, b_1, b_2 and b_3. Tabulate your results.

2. Calculate the percentage carry-over for each cam.

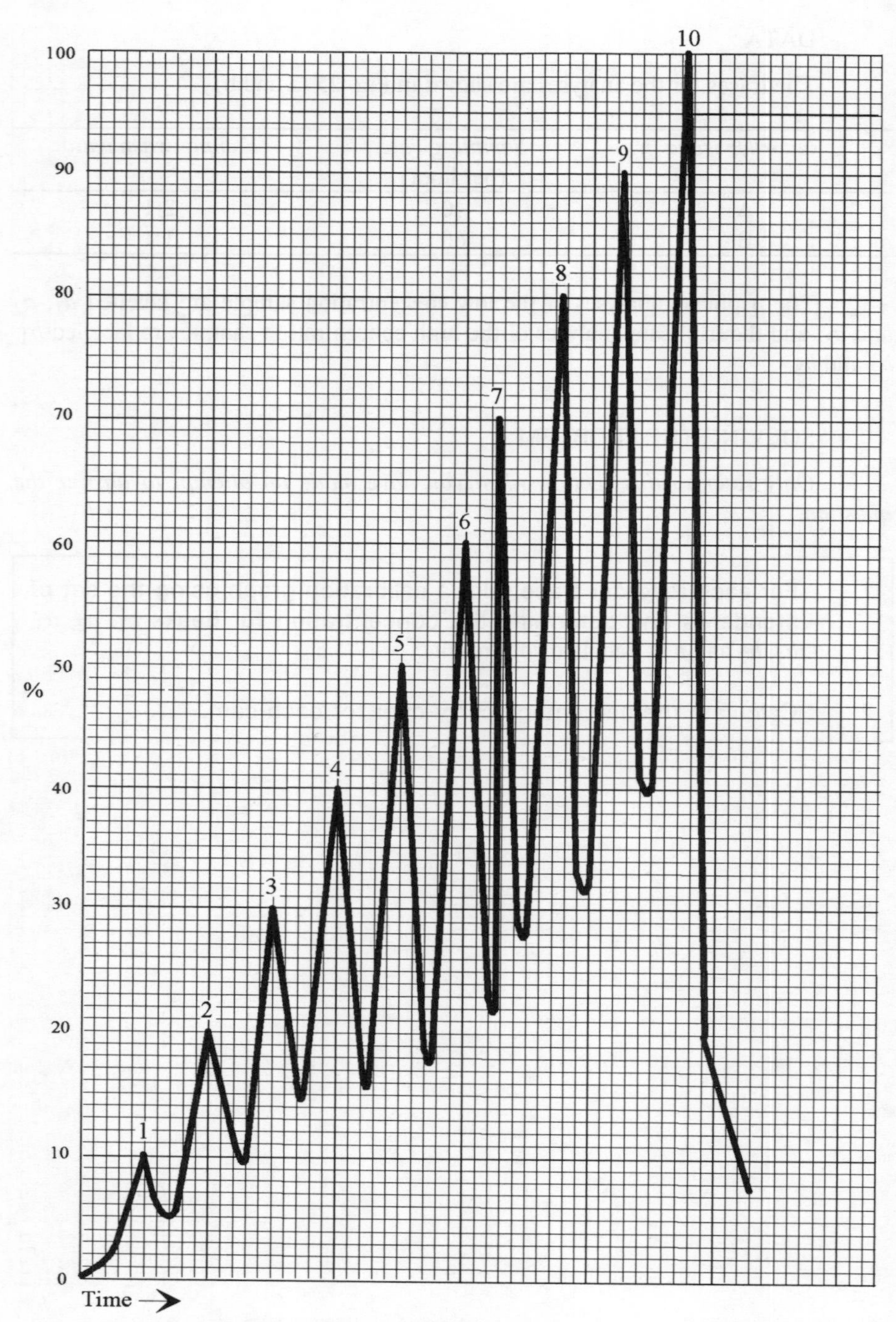

Fig. 25 Autoanalyser trace for (a) (i) sample:wash time ratio 2:1. The concentration of each glucose standard in mmol l^{-1} is indicated by the number above the peak.

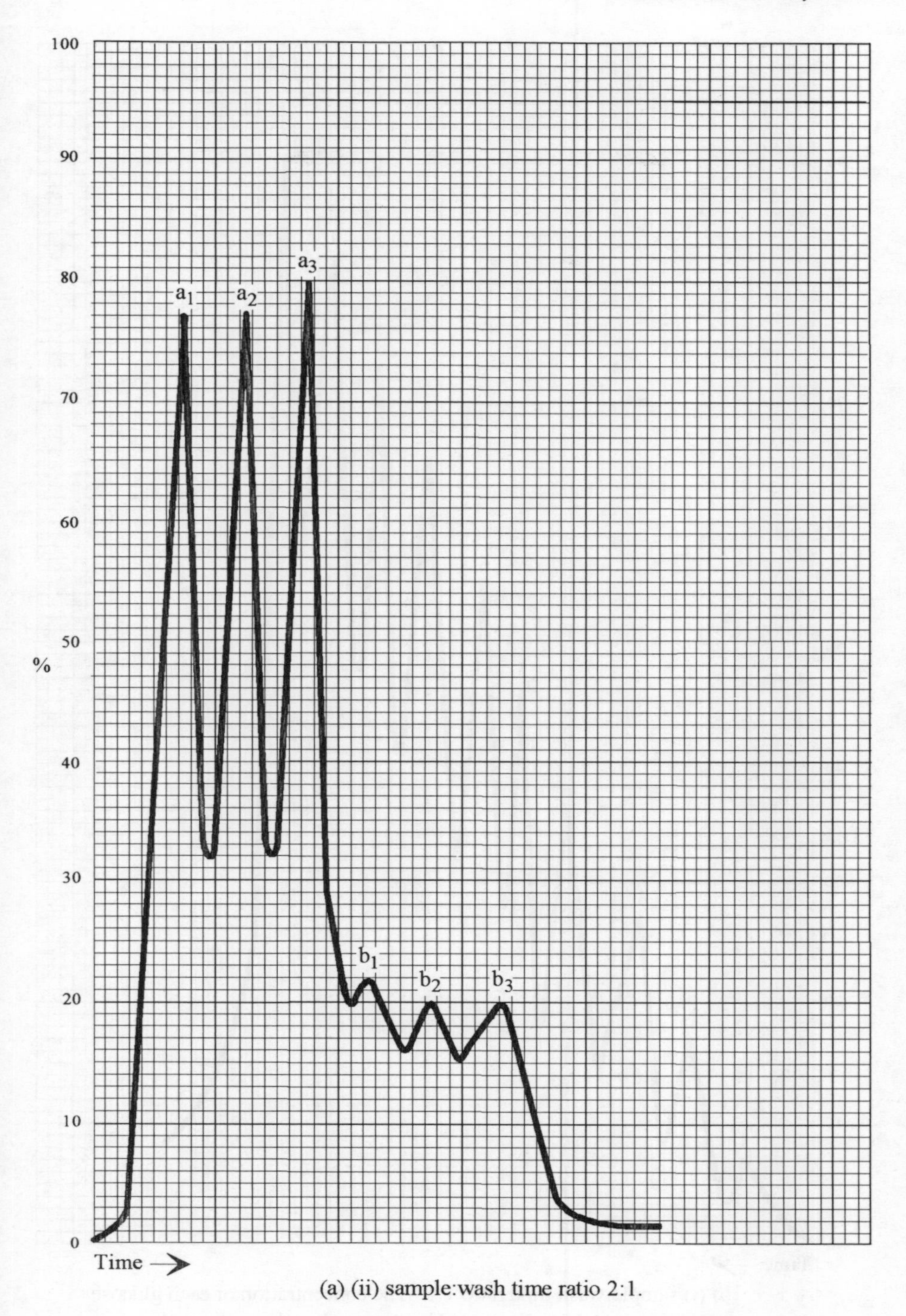

(a) (ii) sample:wash time ratio 2:1.

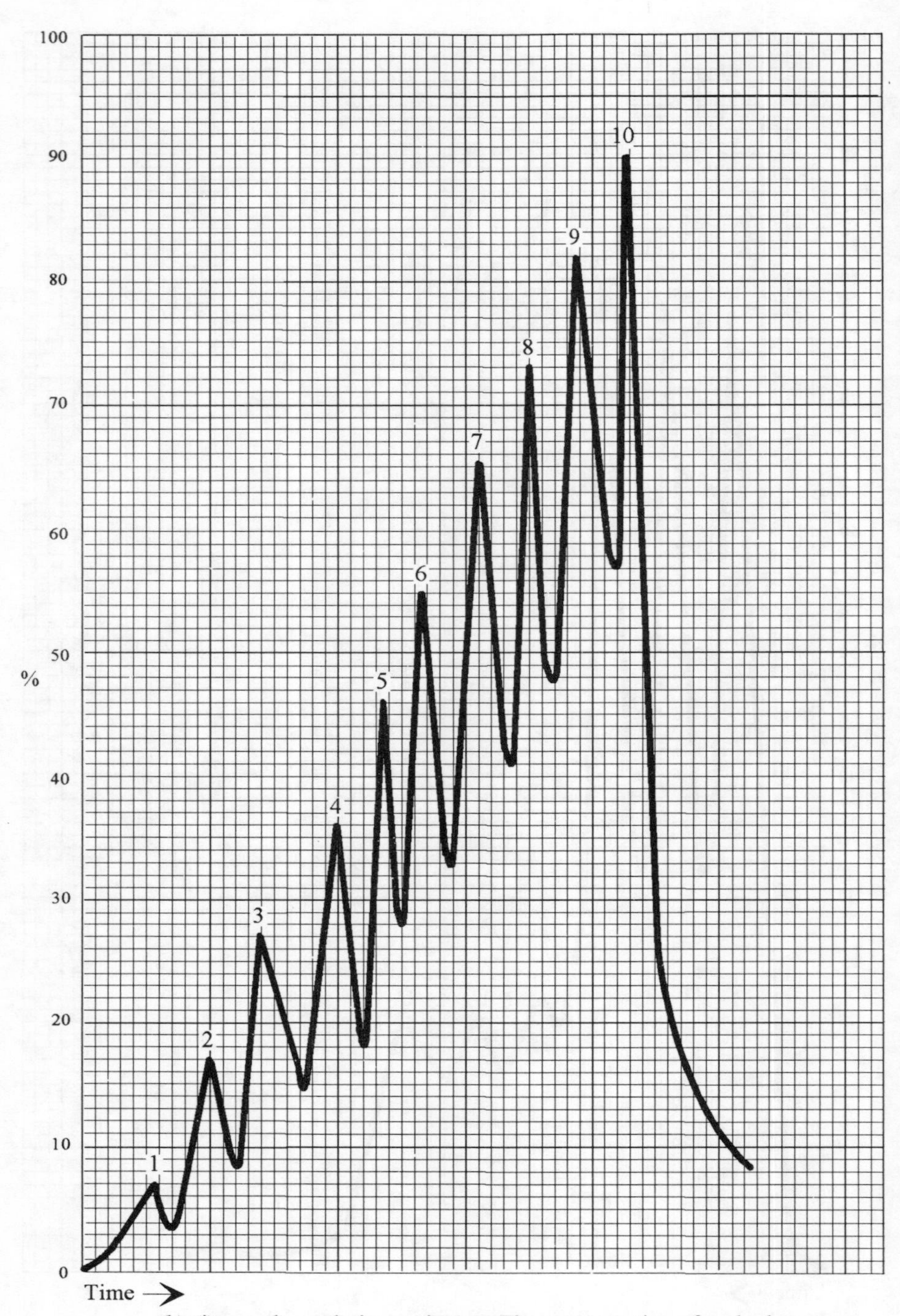

(b) (i) sample:wash time ratio 1:1. The concentration of each glucose standard in mmol l^{-1} is indicated by the number above the peak.

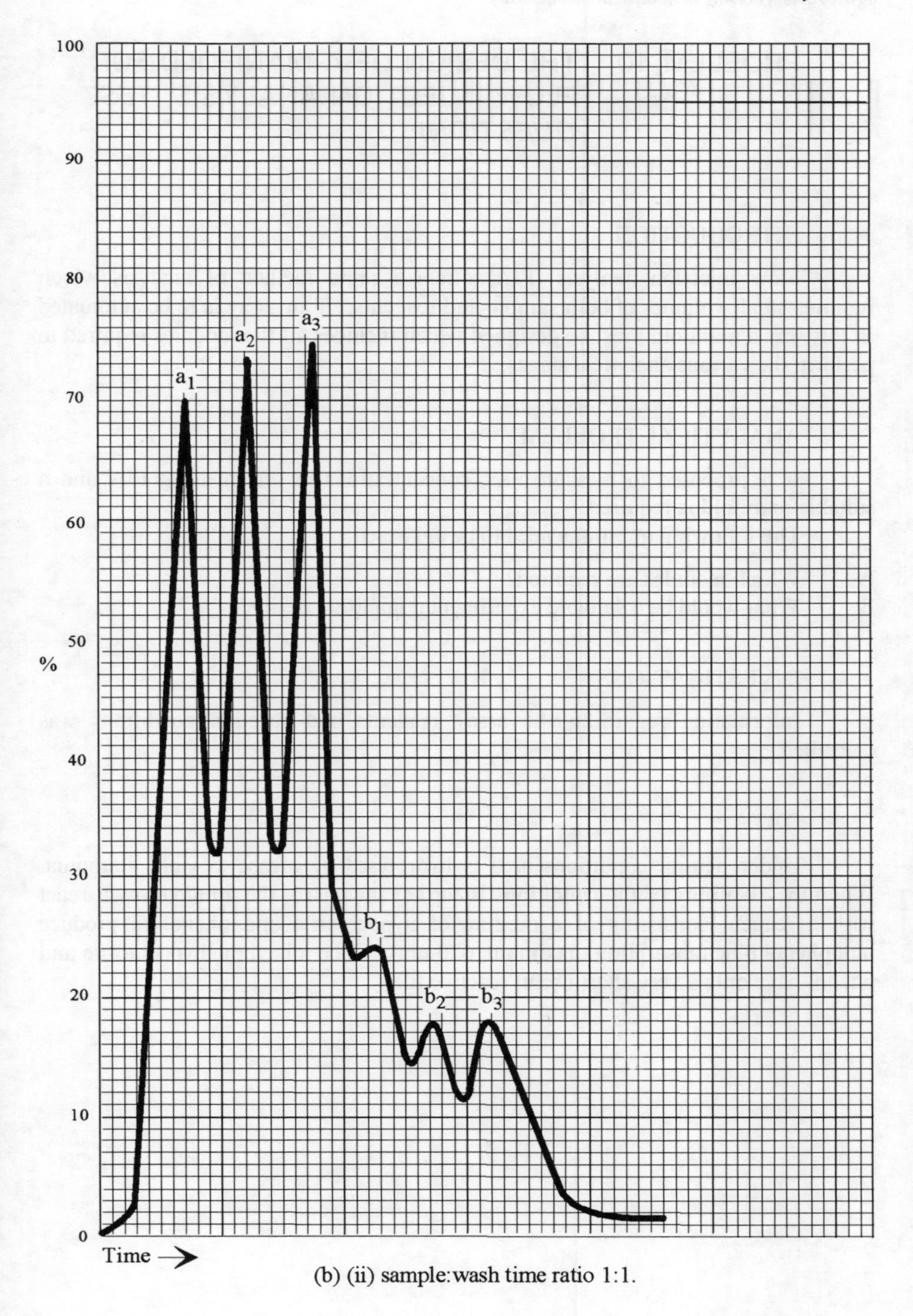

(b) (ii) sample:wash time ratio 1:1.

PROBLEM 40 THE DESIGN OF A MANIFOLD FOR FLOW INJECTION ANALYSIS *(TEST QUESTION)*

INTRODUCTION

Flow injection analysis (FIA) offers a rapid method of analysis which requires small volumes of both sample and reagents. If an assay is to be automated in this way a manifold must be designed which includes all the modules required to perform the various analytical stages.

ANALYTICAL PROBLEM

It is proposed to measure urea concentrations in serum using FIA and a suitable manifold is required.

The following questions need to be answered.

A Which modules are required?
B What would be a suitable experimental manifold design?

INVESTIGATIONS

Information regarding the assay reagents and reaction conditions was consulted.

DATA

Under neutral conditions urea is hydrolysed by urease to yield ammonia. When the alkalinity of the conditions is further increased, the ammonia will react with a reagent consisting of a mixture of hypochlorite and phenol to produce indophenol blue (absorption maximum, 620 nm). Once mixed the hypochlorite and phenol have only a very short useful reactive life.

ANSWERS

PROBLEM 1 A COMPARISON OF THE ANALYTICAL PERFORMANCE OF TWO METHODS

	Method	
	A	B
Mean (mg l^{-1})	13.44	15.25
Standard Deviation (mg l^{-1})	1.50	3.06
Coefficient of variation (V)	11.2%	20.0%
Slope of calibration graph (absorbance mg^{-1})	0.024	0.017
F value	4.12 (critical value 3.05)	
t value	1.56 (critical value 2.10)	

DISCUSSION

The F test suggests that there is a significant difference between the precision of the two methods. The values for V are higher than would be desirable for many analytical methods.

The t test suggests that there is no significant difference between the relative accuracy of the two methods, i.e. there is no significant bias between them.

It would be necessary, in practice, to assess the precision at different concentrations of the analyte.

The calibration graph for method A has the greatest slope indicating a greater degree of sensitivity.

PROBLEM 2 THE USE OF QUALITY CONTROL CHARTS

The Shewart plot indicates that the method is out of control for batch 6. This is not obvious from the Cusum plot which is not suitable for giving immediate information regarding the validity of a particular set of results. It does, however, give valuable information about a change in accuracy, but this is retrospective and several results are needed to demonstrate the change.

The slope of the Cusum plot between batches 1 and 10 indicates that the calculated mean value which was used to set up the chart was slightly incorrect. The change in slope from day 11 onwards shows a change in accuracy of the method. This can be observed as a 'shift' on the Shewart plot.

PROBLEM 3 AN INVESTIGATION INTO THE SPECIFICITY OF AN IMMUNOASSAY

Analyte (ng ml^{-1})	*Reaction rate* (mean)	*Per cent Bound*	*Apparent morphine concentration*	*Per cent Cross reaction*
Morphine				
0	0.540	100	-	-
50	0.387	71.7	-	-
150	0.237	43.9	-	-
300	0.167	30.9	-	-
600	0.120	22.2	-	-
Morphine-6-glucuronide				
100	0.320	59.3	85	85
300	0.200	37.0	215	72
500	0.155	28.7	350	70
1000	0.110	20.4	-	-
Codeine				
100	0.255	47.2	135	135
300	0.165	30.6	310	103
500	0.120	22.2	600	120
1000	0.080	14.8	-	-

DISCUSSION

The assay is quantitative for morphine but the assay is not specific for morphine because both morphine-6-glucuronide and codeine show significant cross-reaction (Fig. 26).

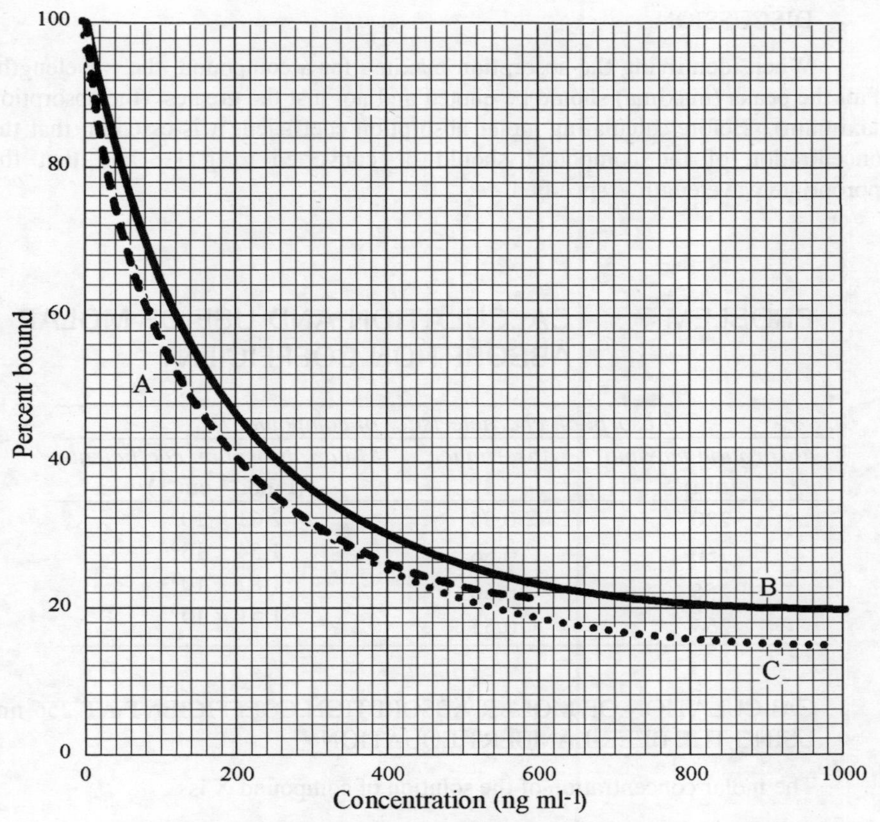

Fig. 26 Calibration graphs for the determination of morphine (A), morphine-6-glucuronide (B) and codeine (C).

PROBLEM 6 AN INVESTIGATION OF THE ABSORPTION AND FLUORESCENT CHARACTERISTICS OF A COMPOUND

Absorption maxima (nm)	*Molar absorption coefficient* ($l\ mol^{-1}\ cm^{-1}$)
A 290	2.2×10^4
B 375	7.6×10^3
C 460	1.0×10^4

Excitation wavelength 470 nm
Emission wavelength 515 nm

DISCUSSION

When identifying the absorption maxima for a compound, the wavelengths of all the peaks (maxima) should be quoted and not just the greatest (the absorption maximum). Before calculating molar absorption coefficient it is essential that the concentration of the compound should be converted to g l^{-1} and that the appropriate wavelength is specified.

PROBLEM 7 CALCULATION AND USE OF MOLAR ABSORPTION COEFFICIENT

Absorption characteristics of compound X

Absorption maxima (nm)	*Absorbance*	*Molar absorption coefficient* (l mol^{-1} cm^{-1})
220	0.95	2.61×10^3
250	0.56	1.54×10^3
256	0.66	1.82×10^3
263	0.55	1.51×10^3

CALCULATION OF MOLAR ABSORPTION COEFFICIENT AT 256 nm USING THE BEER-LAMBERT EQUATION

The molar concentration of the solution of compound X is

$$\frac{0.08}{220} = 3.64 \times 10^{-4} \text{ mol l}^{-1}$$

$$\varepsilon = \frac{A}{cl}$$

$$= \frac{0.66}{3.64 \times 10^{-4}}$$

$$= 1.82 \times 10^3 \text{ l mol}^{-1} \text{ cm}^{-1}$$

CALCULATION OF THE CONCENTRATION OF THE UNKNOWN SAMPLE

The concentration of the unknown sample is:

$$\frac{0.95}{1.82 \times 10^3} = 0.522 \times 10^{-3} \text{ mol l}^{-1}$$

PROBLEM 10 THE DETERMINATION OF OPTIMAL INSTRUMENT SETTINGS FOR ATOMIC ABSORPTION SPECTROSCOPY

DISCUSSION

Optimum sensitivity in this instance is provided when the fuel to oxidant ratio is 2:1, i.e. a reducing flame. It would also be necessary to assess whereabouts in the flame the optimal conditions prevailed, i.e. determine the optimal burner height setting.

The data on p. 24 demonstrates that phosphate suppresses the response due to calcium. This is due to the formation of a thermally stable calcium phosphate complex. The calcium can be released from the phosphate by the presence of lanthanum, which binds the phosphate group more strongly than the calcium and leaves the calcium ions free in the flame.

PROBLEM 12 THE DETERMINATION OF OPTIMAL GAS FLOW IN GLC

Chromatogram	*Retention distance*	$W_{1/2}$	N	HETP
A	21.0	4.0	152	3.29
B	23.0	4.0	183	2.73
C	26.5	4.0	243	2.06
D	32.0	4.0	354	1.41
E	40.0	6.0	246	2.03
F	44.0	8.0	167	2.99

DISCUSSION

A gas flow rate of 2.5 ml min^{-1} is optimal because it gives a minimum value for HETP of 1.41.

These investigations require considerable care and in practice it is advisable to use a high chart speed to give broad peaks and long retention distances to enable half peak width to be measured with some degree of precision.

PROBLEM 13 A STUDY INTO THE EFFECT OF COLUMN TEMPERATURE ON SEPARATION BY GLC

DISCUSSION

Temperature gradient separation is the most suitable for quantitative analysis as is shown by good baseline resolution and good regular separation of the peaks. Low temperature separation produces significant peak broadening and excessive analysis time.

Isothermal separation gives a linear relationship with carbon number and separation at 130 °C would probably be the most suitable for the determination of C-number.

PROBLEM 14 QUANTITATIVE ANALYSIS BY GLC USING A CALIBRATION GRAPH AND AN INTERNAL STANDARD

Sample	*Ethanol concentration* ($g\ l^{-1}$)	*Peak height*		E/P ratio
		Ethanol	Propanol	
Ethanol	5.0	17.5	31.0	0.56
	7.5	26.0	32.0	0.81
	10.0	32.0	30.0	1.07
	12.5	41.5	32.0	1.30
	15.0	45.5	29.0	1.53
	17.5	61.5	32.5	1.89
	20.0	64.0	30.5	2.10
A	(16.0)	51.5	30.5	1.69
B	(11.3)	35.0	29.0	1.21
C	(11.0)	34.0	28.5	1.19
D	(6.0)	18.0	28.0	0.64
E	(5.7)	16.0	26.0	0.62

DISCUSSION

The calculated values for the samples must be corrected for the dilution effect, i.e. the value for wine must be multiplied by 6, and the value for the beer by 3, etc. (Fig. 27).

The value of calibrating an internal standard in this manner is only applicable if the test analyte is available in a pure form, but it does offer a more reliable method of quantitation than relying on a single response factor for a wide range of analyte concentrations.

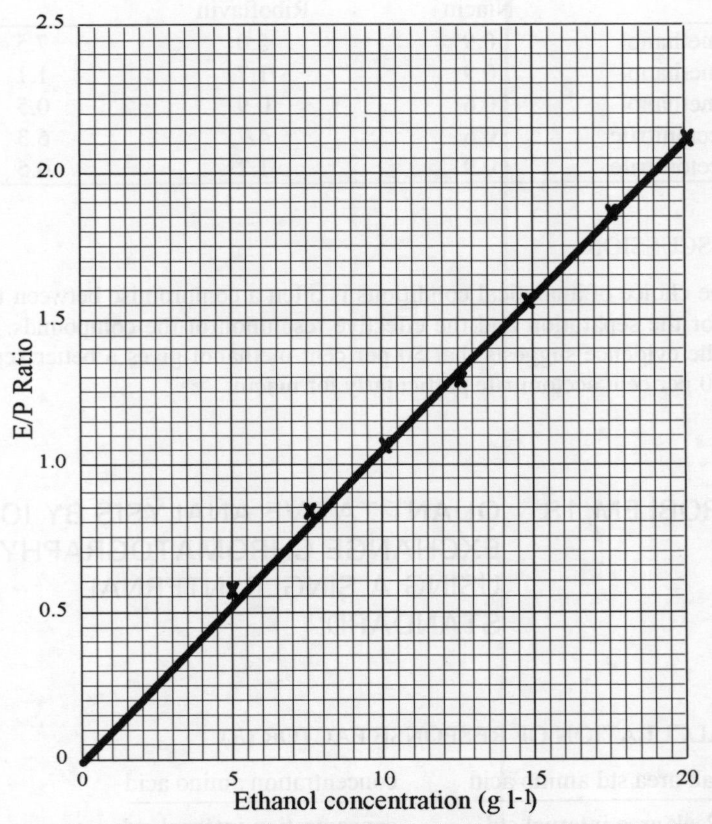

Fig. 27 Calibration graph for the determination of ethanol by GLC.

PROBLEM 17 THE SELECTION OF A MOBILE PHASE FOR REVERSE PHASE HPLC

Solvent	*Retention distance*		*Resolution index*
	Niacin	Riboflavin	
25% methanol	0.9	5.0	7.5
50% methanol	0.9	1.7	1.1
75% methanol	0.6	0.9	0.5
10% acetonitrile	0.6	4.7	6.3
20% acetonitrile	0.9	1.2	0.5

DISCUSSION

The choice of analytical conditions is often a compromise between the time required for the separation and the effective resolution of the compounds. In this example the evidence suggests that 50 per cent methanol gives a better separation than the 10 per cent acetonitrile particularly for niacin.

PROBLEM 18 QUANTITATIVE ANALYSIS BY ION-EXCHANGE CHROMATOGRAPHY USING A SINGLE INTERNAL STANDARD

CALCULATION OF RESPONSE FACTOR (R)

$$\frac{\text{Peak area std amino acid}}{\text{Peak area internal std}} \times R = \frac{\text{concentration amino acid}}{\text{concentration internal std}}$$

Results

Histidine $R = \dfrac{22}{30} = 0.73$

Phenylalanine = 1.1

Tyrosine = 1.2

CALCULATION OF AMINO ACID CONCENTRATION

$$\text{Histidine} = \frac{18}{16} \times 0.73 \times 2 \times \frac{60}{40} \times \frac{20}{60}$$

$$= 0.82 \text{ mol } l^{-1}$$

Phenylalanine = 1.36 mol l^{-1}

Tyrosine = 0.75 mol l^{-1}

PROBLEM 19 THE DETERMINATION OF RELATIVE MOLECULAR MASS USING GEL PERMEATION CHROMATOGRAPHY

Protein	V_e	V_e/V_0	log_{10} RMM
Alcohol dehydrogenase	3.5	2.09	5.176
Bovine albumin	4.0	2.35	4.819
α-amylase	3.3	1.96	5.301
Cytochrome c	5.1	3.00	4.093
Unknown enzyme	3.6	2.11	5.150

DISCUSSION

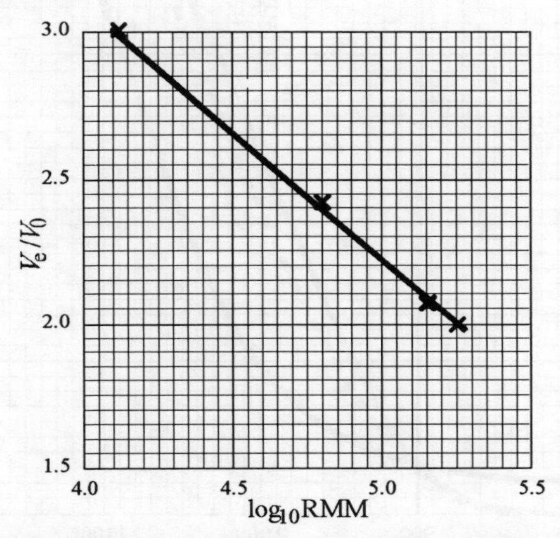

Fig. 28 Determination of relative molecular mass by gel permeation chromatography

The elution distance for Blue Dextran of 4.5 gives the value for V_0. A graph of log RMM against V_e/V_0 gives a straight line and reading off the value for the unknown enzyme gives a value for the RMM of 1.42×10^5 da (Fig. 28).

PROBLEM 20 SELECTION OF THE OPERATING VOLTAGE FOR ELECTROCHEMICAL DETECTION IN HPLC

DISCUSSION

1150 mV.

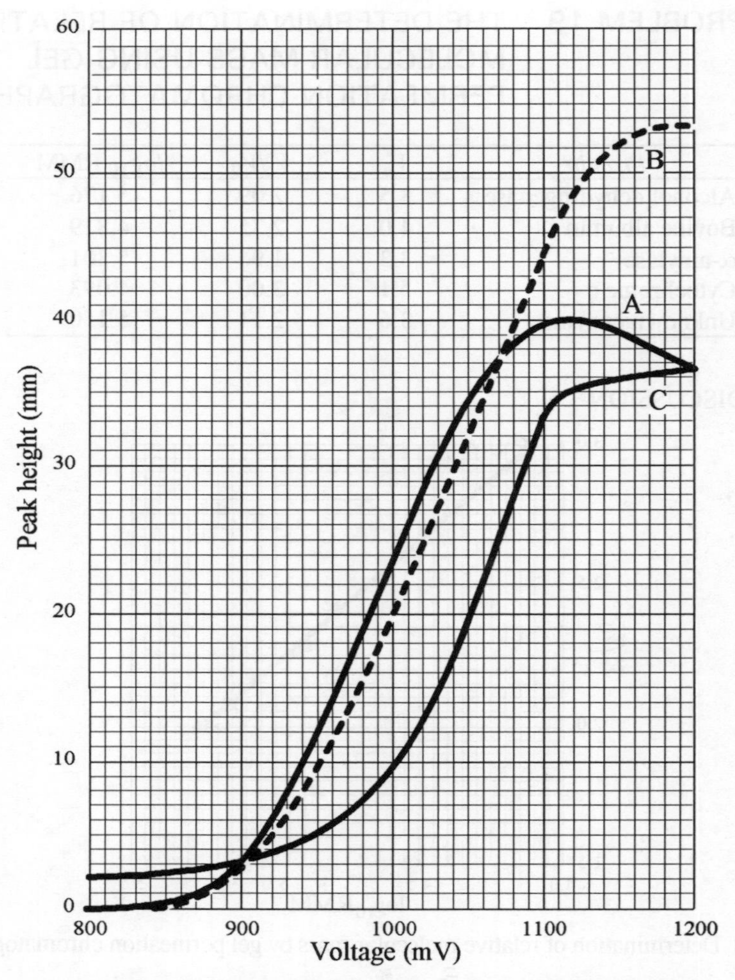

Fig. 29 Determination of optimum operating voltage for an electrochemical detector in HPLC. (A) p-aminobenzoic acid, (B) m-hydroxybenzoic acid, (C) 0-aminobenzoic acid.

PROBLEM 23 DETERMINATION OF pK_a VALUES BY TITRATION

A The trace shows three inflections relating to three ionizable groups.

B pK_a values = 2.0, 6.3 and 9.5 (refer to *Analytical Biochemistry*, § 10.1.3).

C Contains one acidic group and two basic groups.

D Volume of NaOH required for ONE ionizable group = 1.2ml.

$$\text{Molarity of the sample} = 3 \times 1.2 \times \frac{0.1}{1000}$$

$$= 3.6 \times 10^{-4} \text{ mol l}^{-1}$$

$$\text{RMM of the sample} = \frac{0.05}{\text{molarity}}$$

$$= 139$$

PROBLEM 24 THE SELECTION OF SEPARATION CONDITIONS FOR ION-EXCHANGE CHROMATOGRAPHY

DISCUSSION

An anion exchange resin would be the most appropriate because the p*I* values for the test proteins lie between pH 5–7, and at a pH of 7 or greater all the proteins would carry a negative charge. The pH range of 5–7 lies within the ionization range of the CAX resin.

A gradient starting at pH 7 and falling to pH 4 would result in all the proteins being in the undissociated form in sequence, and hence being eluted from the column in sequence. The choice of the same medium, but using a pH 4 to start the separation, would result in none of the proteins showing any affinity for the ion-exchange medium.

PROBLEM 25 THE DETERMINATION OF RELATIVE MOLECULAR MASS USING SDS ELECTROPHORESIS

Sample	*RMM*	log_{10}	*Distance*
Standards	94,000	4.9731	4
	67,000	4.8261	10
	43,000	4.6335	16
	30,000	4.4771	25
	20,000	4.3032	39
	14,400	4.1584	55
Sample X	33,800	4.5300	20
Enzyme	16,800	4.2250	48
Enzyme treated X	-	-	20
	-	-	48

DISCUSSION

The results show that the protein is unaffected by enzymatic treatment (Fig. 30).

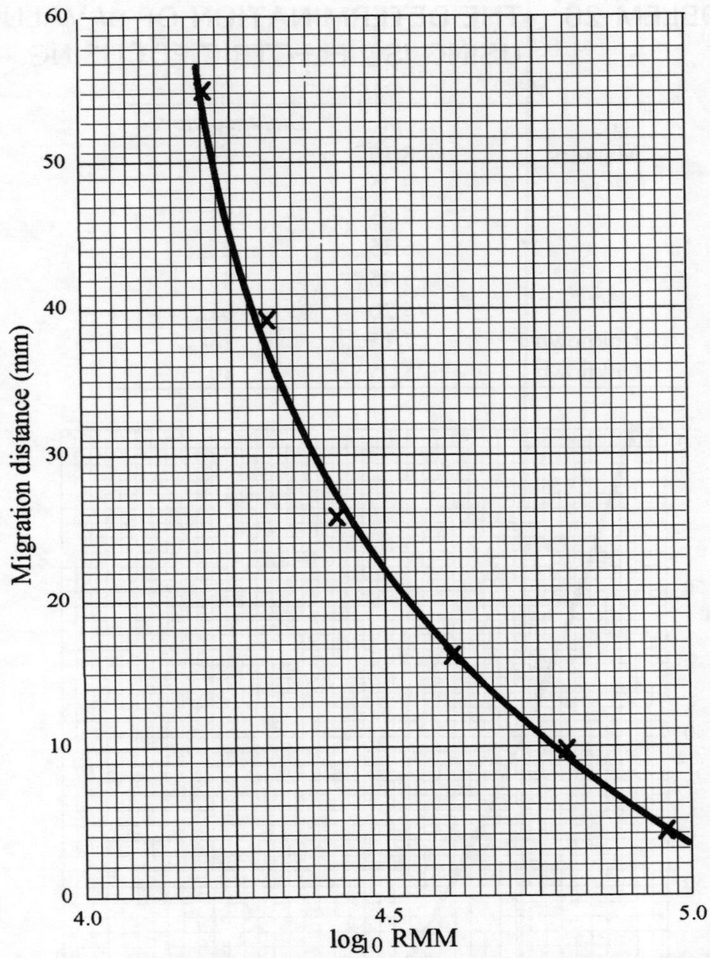

Fig. 30 Determination of relative molecular mass by SDS electrophoresis.

PROBLEM 26 THE DETERMINATION OF p*I* VALUES USING ISO-ELECTRIC FOCUSING

	p*I*	*Distance moved*
Markers	8.15	5
	7.35	15
	6.85	23
	6.55	27
	5.85	45
	5.20	61
Unknown 1	7.15	18
Unknown 2	5.95	40

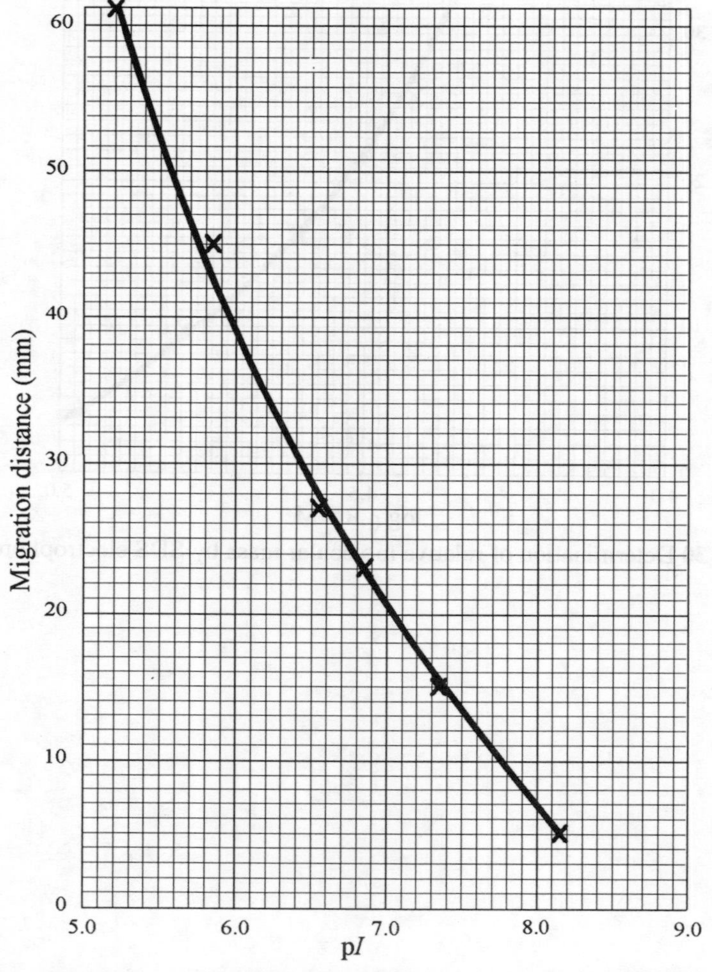

Fig. 31 Determination of p*I* values by iso-electric focusing.

PROBLEM 29 THE USE OF A RADIOACTIVE TRACER TO ASSESS THE EFFICIENCY OF AN EXTRACTION PROCEDURE

A $$\frac{\text{Sample B} - \text{background count}}{\text{Sample A} - \text{background count}} \times 100\% = 1.6\%$$

PROBLEM 30 THE USE OF ISOTOPE DILUTION ANALYSIS FOR QUANTITATION

$$\text{Test}_{(\text{conc})} = \text{Tracer}_{(\text{conc})} \times \left(\frac{\text{Isotope specific activity}}{\text{test specific activity}} - 1 \right)$$

$$= 1000 \times \left(\frac{2450 - 15}{1340 - 15} - 1 \right)$$

$$= 837 \text{ pg ml}^{-1}$$

PROBLEM 33 THE OPTIMIZATION OF ASSAY CONDITIONS OF A COUPLED KINETIC ENZYME ASSAY

DISCUSSION

A The graph of enzyme activity showed that a pH 8.5 would give maximum sensitivity for the test enzyme RAO. In order to compensate for the reduced activity of the indicator enzyme at that pH, additional GDH could be added.

B A substrate concentration of at least ten times the Km value is suggested to give maximal activity. The Km for RAO is 1.05×10^{-6} mol l^{-1} so a concentration of at least 100 times greater would be suitable (Figure 32).

C Both enzymes are inhibited by the co-factors required by the other enzyme, RAO by ADP and GDH by FAD. The inhibitory effects of one are relieved by an increased concentration of the co-factor. Maximum activity of both enzymes would be achieved in 1.0 mmol l^{-1} ADP and 0.1 mmol l^{-1} FAD.

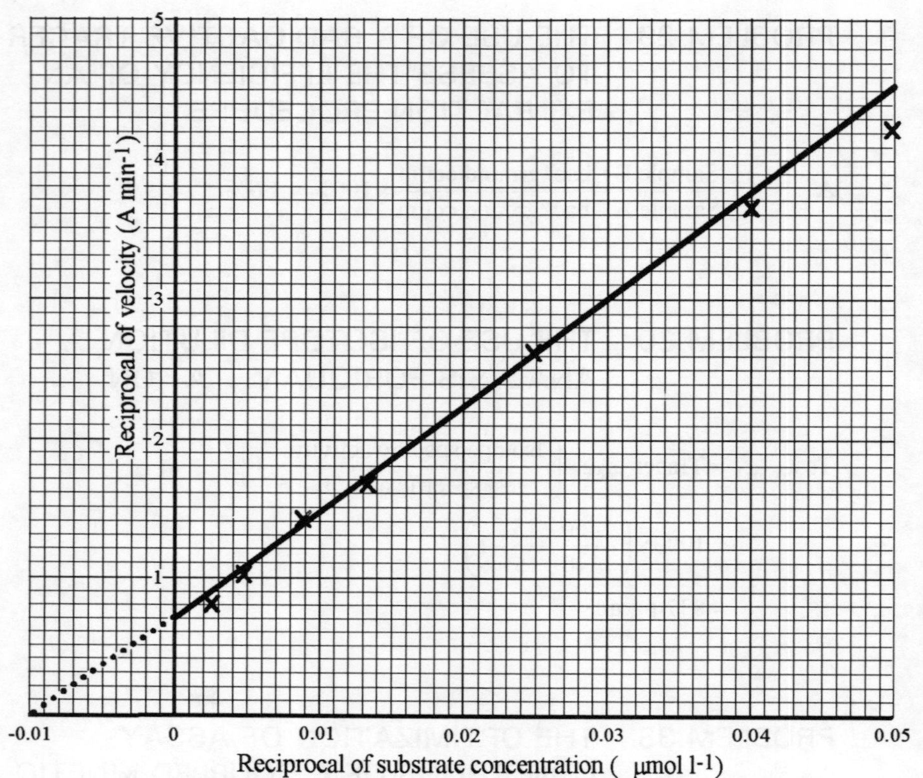

Fig. 32 Determination of K_m for the enzyme R-amine oxidase.

PROBLEM 34 THE DEVELOPMENT OF A FIXED TIME ENZYME ASSAY

DISCUSSION

A From the graph of the table given on p. 134, the reaction is approaching completion at about 15 min, and probably a time a little shorter than this might be suitable.

B There is a linear relationship up to the maximum activity of the sample for reaction times of 2 and 5 min but the sensitivity of the method is low. At 10 min the sensitivity is greater but the linear relationship is only true for an enzyme activity equal to over 90 per cent of the sample. The range of the method is as low as about 50 per cent of the sample for a reaction time of 18 min. The choice of reaction time depends to a large extent on the analytical range required but if the maximum activity likely to be encountered is the same as the sample analysed, a reaction time of 10 min would be suitable.

PROBLEM 35 THE CALCULATION OF ENZYME ACTIVITY

Kinetic assay

Initial velocity as absorbance change per minute = 0.08.

$$\text{Enzyme activity} = \frac{0.08}{5.0 \times 10^3} \times \frac{3}{1000} \times \frac{1}{60} \times \frac{1}{0.1} \text{ katal ml}^{-1}$$

Fixed time assay

$$\text{Concentration of product formed} = \frac{0.35}{1.80} \times \frac{0.5}{1000} \text{ mmol l}^{-1}$$

$$\text{Enzyme activity} = \frac{0.35}{1.80} \times \frac{0.5}{1000} \times \frac{5}{1000} \times \frac{1}{10 \times 60} \times \frac{1}{0.1} \text{ katal ml}^{-1}$$

DISCUSSION

The confusion that arises in attempting to calculate enzyme activity can be reduced if the steps suggested are undertaken sequentially.

1. Calculate the change in concentration in 1 s.
2. Calculate the actual amount of product in the assay volume.
3. Correct for the volume of sample used in the assay.

PROBLEM 38 THE DESIGN OF A MANIFOLD AND FLOW DIAGRAM FOR AIR-SEGMENTED CONTINUOUS FLOW ANALYSIS

See Figure 33.

DISCUSSION

The quality of any peaks obtained when the system is tested using glucose solutions will determine how to proceed. At this stage modifications may be required, e.g. alteration of reagent concentration, addition of extra mixing coils to increase the reaction time, etc. The sampling rate and carry-over then need to be assessed.

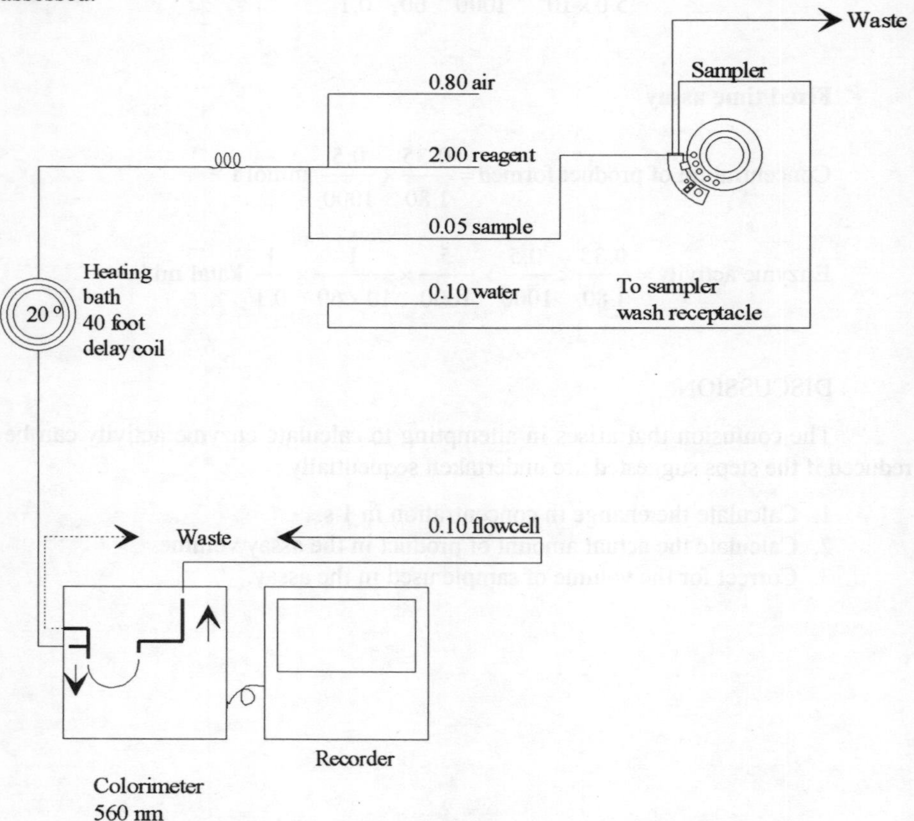

Fig. 33 Flow diagram for the determination of glucose by air-segmented continuous flow analysis.

PROBLEM 39 THE ASSESSMENT OF CARRY-OVER IN AIR-SEGMENTED CONTINUOUS FLOW ANALYSIS

Cam A $$\text{Carry-over (\%)} = \frac{2.2-2.0}{8.0-2.0} \times 100 = 3.3\%$$

Cam B $$\text{Carry-over (\%)} = \frac{2.7-2.1}{8.2-2.2} \times 100 = 10\%$$

DISCUSSION

Only a limited range of cams is available and selection is a compromise between the time for analysis and the carry-over value. Sampler units incorporating timing systems offer greater flexibility.